T0075458

SpringerBriefs in Earth Sciences

SpringerBriefs in Earth Sciences present concise summaries of cutting-edge research and practical applications in all research areas across earth sciences. It publishes peer-reviewed monographs under the editorial supervision of an international advisory board with the aim to publish 8 to 12 weeks after acceptance. Featuring compact volumes of 50 to 125 pages (approx. 20,000–70,000 words), the series covers a range of content from professional to academic such as:

- timely reports of state-of-the art analytical techniques
- bridges between new research results
- snapshots of hot and/or emerging topics
- literature reviews
- in-depth case studies

Briefs will be published as part of Springer's eBook collection, with millions of users worldwide. In addition, Briefs will be available for individual print and electronic purchase. Briefs are characterized by fast, global electronic dissemination, standard publishing contracts, easy-to-use manuscript preparation and formatting guidelines, and expedited production schedules.

Both solicited and unsolicited manuscripts are considered for publication in this series.

More information about this series at http://www.springer.com/series/8897

J. H. L. Voncken

Geology of Coal Deposits of South Limburg, The Netherlands

Including Adjacent German and Belgian Areas

J. H. L. Voncken
Department of Geoscience and Engineering
Faculty of Civil Engineering and
Geosciences
Delft University of Technology
Delft, The Netherlands

ISSN 2191-5369 ISSN 2191-5377 (electronic)
SpringerBriefs in Earth Sciences
ISBN 978-3-030-18285-4 ISBN 978-3-030-18286-1 (eBook)
https://doi.org/10.1007/978-3-030-18286-1

This Springer imprint is published by the registered company Springer Nature Switzerland AG
The registered company address is: Gewerbestrasse 11, 6330 Cham, Switzerland

This book is dedicated to the memory of my father, Johannes Lambertus Voncken (1928–2003).

Preface

This book is inspired by my long-standing interest in Dutch coal mining. This stems from my youth, as my father (J. L. Voncken, 1928–2003) worked as a miner, and later as a shot firer/blaster in the Dutch coal mine Oranje Nassau Mine II. This mine was located in the village of Schaesberg, which is presently part of the municipality of Landgraaf. His stories about underground technology and also the fossils and minerals he sometimes brought home to show us fascinated me. This definitely influenced me in my choice of university study: the study of geology (rather than that of mining, as I was more interested in the history of the Earth, and the formation of rocks and minerals, than in mining).

Having started in 2002—together with a (now former) university colleague—a Web site (off-line since 2017) on Dutch coal mining, I became familiar with the technology (and its development), which had been used from the beginning of the twentieth century until the years 1968–1974, when the mines were, one after another, closed, following the decision of the Dutch Government in 1965 to end all coal mining activities in South Limburg.

When first pondering writing a book about the coal mines, I considered writing a book mainly about the nature of coal and the geology of the area, which then would also include the neighboring German and Belgian mining areas. All these mining areas are in what is, de facto, one large coalfield. This coalfield is, however, dissected by several large, active, faults. There are also already many books in existence about the mines themselves, especially about the closure of the mines, and there are also several semiprofessional and amateur websites devoted to the subject of Dutch coal mining.

The geology of the area has, of course, been investigated in great detail, but most published works date from the 1950s and 1960s. With respect to the formation of the coal deposits, these texts do not take into account the fact that Western Europe has not always been located on the face of the Earth where it is now: the concept of plate tectonics was not considered as a factor in these analyses. Even at the time that I started my study of geology, in the late 1970s, plate tectonics was presented as a probably correct theory, though still not a generally accepted one.

In many studies from the 1950s and 1960s, it was thus postulated that the Earth in the Carboniferous period had a tropical climate throughout its entirety. Modern views consider this assumption to be completely incorrect and point instead to the concept of plate tectonics as a corrective view.

This book does exactly this: It takes into account the important effect of plate tectonics. The formation of the coal deposits took place in a tropical climate, due to the position of Western Europe on the face of the Earth during the Carboniferous period, when it was located in the tropics. Western Europe was part of a former microcontinent, called Avalonia, which, starting on the southern hemisphere, drifted northward over the surface of planet Earth during hundreds of millions of years.

The book starts with an overview of mining activities in the Aachen–South Limburg–Campine area, which had started already during the Middle Ages. In Chap. 2, the nature of coal, the formation of coal, its characterization, and the classification of coal are described. Chapter 3 outlines the wanderings of this part of Western Europe over the surface of the Earth during hundreds of millions of years, starting in the Precambrian, and more in detail from the Silurian period until the Carboniferous period. Chapter 4 describes what happened during the hundreds of millions of years after the Carboniferous period, in which the coal itself was formed. This chapter focuses especially on the Roer Valley Graben, an active graben system that is a western branch of the Rhine Graben rift system. The Rhine Graben rift system is a main feature in the geology of Western Europe.

I have placed the subject of the book therefore in the context of the geological development of the area through time, using up-to-date theories of plate tectonics.

In the South Limburg area, in the Miocene epoch, also important quartz sand deposits were formed, which have been mined from the beginning of the twentieth century until the present day. Locally occurring small deposits of lignite (i.e., brown coal) have been exploited in the past, but mining ceased in the 1950s.

Delft, The Netherlands J. H. L. Voncken

Acknowledgements

I would like to express my thanks to Dr. M. W. N. Buxton, Head of the section Resource Engineering at Delft University, who allowed me to write this book as part of my work at the university.

I also would like to express my thanks to my colleague Dr. K.-H. A. A. Wolf, in the section of Petrophysics at Delft University, who helped me with the coal petrography section, and who has, in past years, given me a lot of material and literature references on the subject of coal in general.

I would also like to thank Dr. Janice Rossen, who edited the book manuscript.

Contents

Chapter 1
Introduction—History of Coal Mining in South Limburg/Aachen Area/Campine Coalfield

Abstract This chapter tells about the history of coal mining in the Dutch area of South Limburg, and in the adjacent German (Aachen) and Belgian (Campine) areas. Coal mining in this region of Europe started in the Middle Ages. Coal mining on an industrial level started in the German and Belgian areas in the middle of the 19th Century and beginning of the 20th Century respectively. In The Netherlands it began in the 1890s, and into the beginning of the 20th Century.

Keywords History of coal mining · South Limburg · Aachen area · Campine area · Shaft sinking · Artificial Ground Freezing (AGF)

1.1 The South Limburg Area and the Aachen-South-Limburg-Campine Coalfield

South Limburg (Fig. 1.1) is well known as a picturesque part of The Netherlands, especially as it is not flat, as is most of the country, but rather has an undulating topography, consisting of a major plateau area (Kunrade Plateau and Margraten Plateau), cut by rivulets (Geul, Gulp) and bordered in the West by the Meuse river. For the rest, it consists mostly of hilly terrain, which is cut and formed by several small rivers and streams. The northern part, and especially the north-eastern part of South Limburg, is also well known as being an area of intense coal mining activity from 1899 until 1974. Coal mined here was used in the whole of The Netherlands as heating material, and also partly in order to produce coke and gas. The gas was separated into its components, and these were used for various ends. Figure 1.2 shows the coalfields of Western Continental Europe.

The closure of the mines, which because of diverse reasons had become uneconomical, lasted from 1967 until 1974, and caused major unemployment, as well as the economic collapse of the region. This subject is extensively covered in many books and historical and sociological studies (e.g. Messing 1988).

J. H. L. Voncken, *Geology of Coal Deposits of South Limburg, The Netherlands*, SpringerBriefs in Earth Sciences, https://doi.org/10.1007/978-3-030-18286-1_1

Fig. 1.1 South Limburg in The Netherlands (*Source* http://www.landkaartnederland.net/, adapted). The Aachen area is located directly to the east, and the Campine area directly to the west. See also Fig. 1.2

The geology of the coal deposits has in the past been studied in detail, with respect to the stratigraphy of the coal-layers, and with respect to major and minor faults and folds, disturbing the stratigraphy (e.g. Patijn and Kimpe 1961; Wrede and Zeller 1988). Many studies, however, were also made by coal mining companies themselves, and are not always available to the public.

1.2 Coal Mining in South Limburg

1.2.1 Introduction

Coal mining in South Limburg actually started in the Middle Ages. Coal was found in the region of Kerkrade (the Land of Rode) since medieval times. The center of the coal mining in the region was at first located in the valley of the Worm (German: Wurm) River, which had cut deeply into the surrounding land, and which also cut

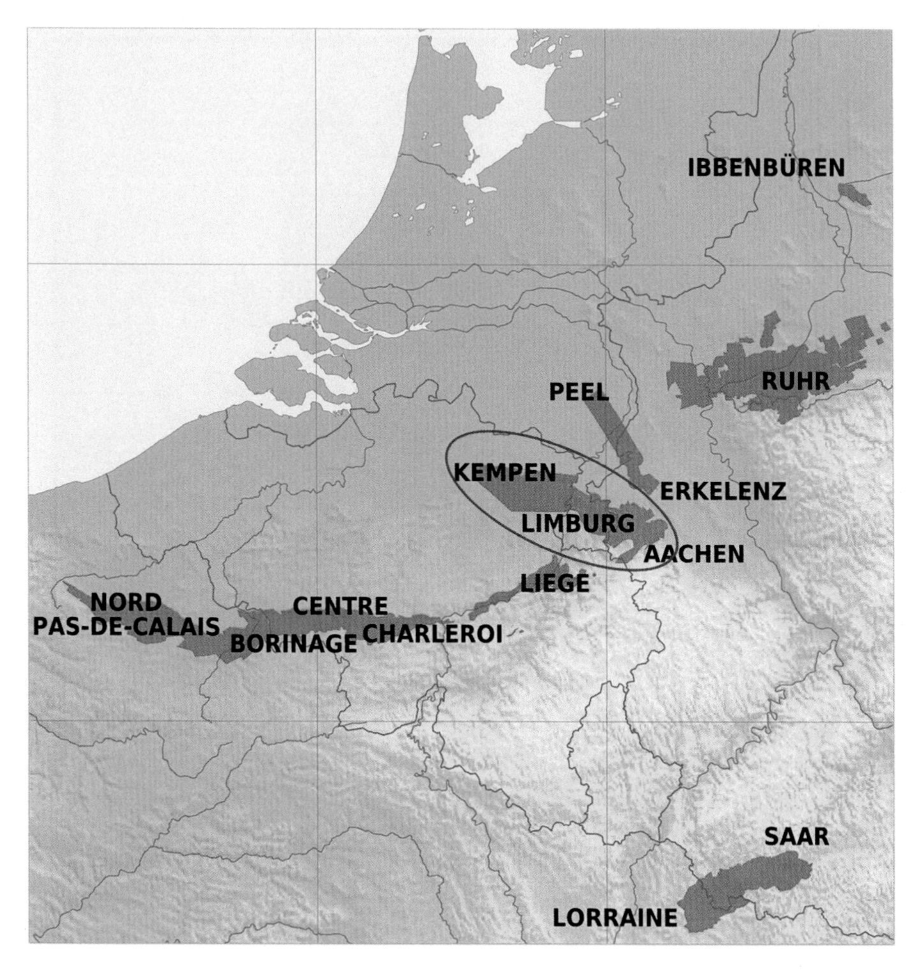

Fig. 1.2 The coalfields of Western Continental Europe. In the ellipse, the location is given of the coalfield that is discussed in this book. It is one large coalfield, covering area in three countries: Germany, The Netherlands, and Belgium. The Peel coalfield in The Netherlands has never been exploited. "Kempen" (Belgium) is in English referred to as "Campine" (After https://en.wikipedia.org/wiki/List_of_coalfields. Licensed under CC BY-SA 3.0. Original image by H. Erren. Adapted)

through several coal seams. In later times, the abbey Kloosterrade became a centre of the local mining industry, and the abbots of Kloosterrade became wealthy as a result (Voncken and De Jong 2003). Although coal was initially found at the surface, soon small underground mines developed, like those shown in Fig. 1.3.

Fig. 1.3 Shaft sinking and underground mining in the Middle Ages [From Agricola (1556)]

1.2.2 Shaft Sinking

Much of Central Europe is underlain by a series of strata which are heavily water-bearing and as a result are very difficult to sink shafts through. In 1884, in Germany, Friedrich Hermann Poetsch patented the freezing method (Artificial Ground Freezing or AGF) for shaft sinking through heavily water-bearing soil (Poetsch 1884). This system was very popular, and in Germany, France, Poland, The Netherlands, and Belgium over 100 shafts in total were sunk using this method. The method is outlined in brief below.

Artificial Ground Freezing (AGF) is a soil stabilization technique involving the removal of heat from the ground to freeze a soil's water in pores. The method involves a system of pipes consisting of an outer pipe and a concentric inner feed-pipe, through which a chilled coolant (originally calcium chloride brine) is circulated. Schematically, the system is illustrated in Fig. 1.4.

A series of boreholes (every 0.8–2 m) is made in a circle, into which freezing pipes are installed (Fig. 1.4). Coolant liquid cooled to −20 to −40 °C is flushed through the pipes, using pumps. As a result of the constant circulation of cooling liquid through the pipes, the water-bearing soil outside the pipes freezes. Ice wall cylinders are formed around each pipe, which, with persistent cooling, merge to

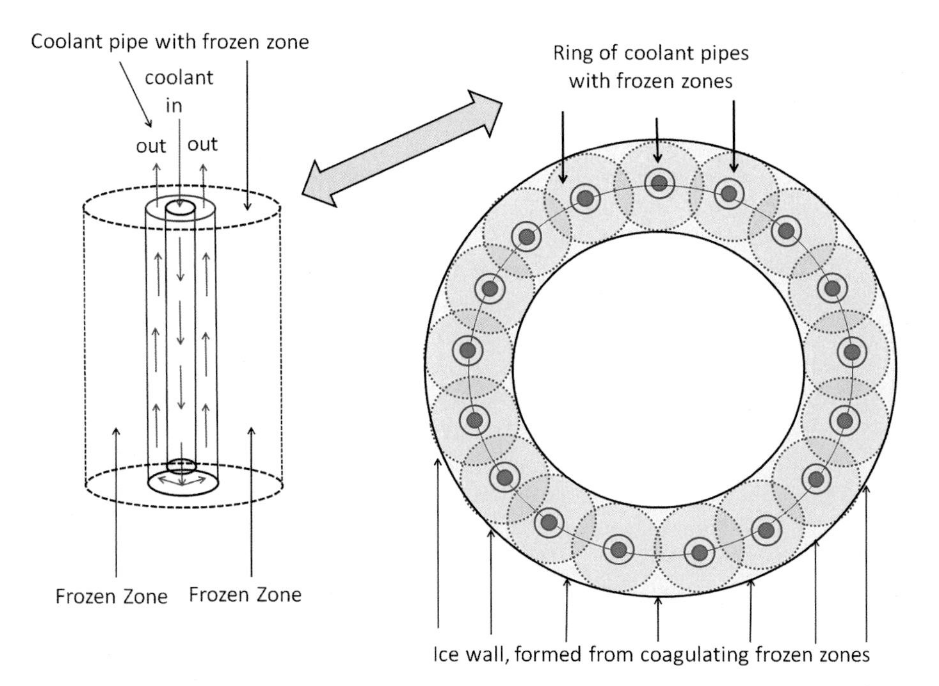

Fig. 1.4 The AGF-process explained. An ice-wall is formed as a result of the coagulation of frozen zones around each pipe assembly. Within this ice-wall, excavation of the shaft can be carried out safely. Own image

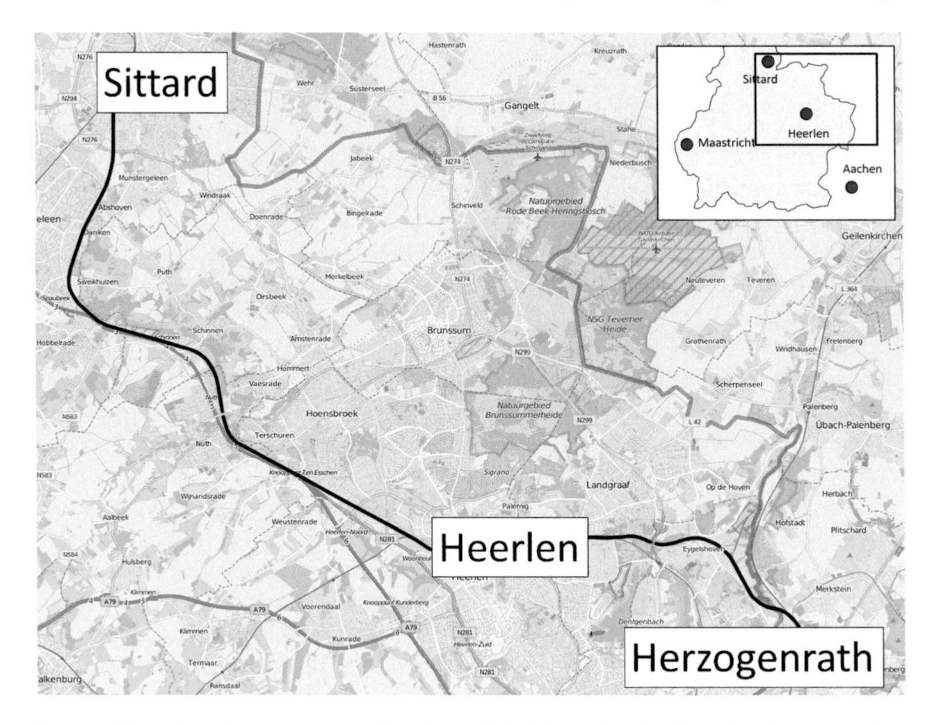

Fig. 1.5 The railway Herzogenrath—Sittard (bold black line), built by Henri Sarolea (Image from Google Maps and Wikipedia, adapted)

form one solid ice cylinder (Fig. 1.4). Mining operations can start in the center of the frozen cylinder, when the right dimensions are reached.

A further development on the AGF technique occurred in France in 1962, when liquid nitrogen was pumped into the freeze pipes, instead of chilled calcium chloride brine. This allows for much faster ground freezing. The liquid nitrogen runs through the freeze pipes and is allowed to evaporate into the atmosphere (Geo-engineer 2017).

Another method to sink shafts was invented by the German mining engineer Friedrich Honigmann (1841–1913), and subsequently perfected by Dutch mining engineer Gerard Jan de Vooys (1902–1979).

This method (called the Honigmann—De Vooys drilling method, or, for short, the drilling method) consisted of drilling a shaft with initially a drill, then applying knives to loosen the material in the soft top soil, and further down a normal drill. The walls of the shaft were protected from collapsing by using drilling mud. This drilling mud consisted of a mushy mixture of water and clay, which has a higher specific weight than water, and as a result keeps the water from outside flowing into the shaft. On the wall of the shaft, a clay layer is formed, closing off the water influx from outside, The rubble created by the drilling is (together with the mud) pumped to the surface by means of a tubing system. In a basin, the mud/water mixture is allowed to settle, allowing the rubble to sink to the bottom. The cleaned mud/water mixture than is reused.

1.3 Coal Mines in South Limburg

1.3.1 *Domaniale Mine*

The Domaniale Mine was the oldest mine in the South Limburg Area. Mining had already started during the Middle Ages. In the second half of the 18th Century, the abbot of the Kloosterrade Abbey appointed several technically skilled people from the coal district of Liège (Belgium) in his service, which led to an increase in coal production and in mining skills. The most important mine in the region flooded almost daily, but the new technicians made use of the water power of the Worm River in order to drive pumps, which kept the mine dry. In 1789, the French Revolution initiated the end of the political power of the abbey. When in the beginning of the 19th Century French troops occupied The Netherlands, they nationalized the mines of the abbey, and called them "Mines Domaniales." When the French departed, the mine fell under the jurisdiction of the new Dutch Kingdom, and was called Domaniale Mine. The Domaniale Mine was closed in 1969.

1.3.2 *Transportation Possibility*

Mining, however, remained still quite small in scale, as the mined coal could not be transported over large distances. Transport had to be done by horse and carriage.

A further boost in mining came in the late 19th Century, when Henri Sarolea[1] (1844–1900), a railway builder from the Dutch East Indies, completed a railway connection between Herzogenrath (Germany), Heerlen, and Sittard (both located in The Netherlands), where the railway connected to the major north-south railway line which ran from the middle of The Netherlands to the very south of the country. This initiated the industrialization of the area around Heerlen, as with the completion of this railway there was now a means of fast transport in the Heerlen area, and Heerlen was connected with other parts of The Netherlands. The Herzogenrath-Sittard railway is shown in Fig. 1.5. Out of gratitude for the achievement of Sarolea, one of the main streets in the centre of Heerlen was named in his honour. The street leads from the railway station to the center of the city.

1.3.3 *Oranje Nassau Mines*

German mining engineer Friedrich Honingmann (1841–1913), son of mining engineer Eduard Honigmann, along with his younger brother Carl (1852–1903), and

[1]The name is pronounced as "Sarólea", with accent on the second syllable.

Table 1.1 Production of the ON-mines

Mine	Years operational	Production (tons)
ON-I	1899–1974	31,978,000
ON-II	1904–1971	34,064,000
ON-III	1914–1973	38,265,000
ON-IV	1927–1966	13,754,000

Source Voncken and De Jong (2003)

Sarolea, started a mining company together. Henri Sarolea took his place on the board (Voncken and De Jong 2003).

In 1893, the Dutch Minister of Transport, Cornelis Lely, gave permission to the new company to exploit the concession Oranje-Nassau (3378 ha). Shortly afterwards, the Honigmann family acquired another concession, named "Carl", with a size of 444 ha. The mining company started in 1894, with the construction of shafts for a mine located at Heerlen, which later was named "Oranje-Nassau Mine I" (Voncken and De Jong (2003).

Officially called "Maatschappij tot Exploitatie van Limburgsche Steenkolenmijnen,[2]" the company became known as "Oranje Nassau Mijnen," or, for short, ON-mines, after the first concession had been obtained: Oranje Nassau (Voncken and De Jong 2003). In due time, 4 ON-mines were opened, named ON-I on through to ON-IV (Voncken and De Jong 2003). The years of operation and total production of these mines is given in Table 1.1.

In 1908, the Honigmann family sold its shares in the Oranje Nassau mines to the French family-owned company De Wendel, ("Les Petits-Fils de François de Wendel et Cie"), who owned steel factories in Lorraine. The Wendel company was interested in coal intended to be used for coke production, and thought that coke could be found in Limburg, although the two then-existing Oranje Nassau mines and the Domaniale mine produced high rank coal for domestic heating, which was not suited for coke production. On the other hand, existing mines in the neighbouring Aachen area (Anna, Nordstern, and Reserve, see Table 1.3) did indeed produce coking coal, so the expectation of the Wendel company in this regard was not unjustified.

With the construction of a third mine at Heerlerheide, in the north of the concession (Oranje-Nassau Mine III, or ON-III, start of construction 1910), and later the addition of a fourth at Heksenberg, the de Wendel company aimed to exploit gas-rich coal seams, but did not find them. Only in more northward and westward positioned concessions, owned by the Dutch State, these coal seams were found abundantly. At Heksenberg, a district of the municipality of Heerlen, the mine ON-IV was constructed from 1910–1927. Originally, the shaft was intended only as a ventilation shaft of the ON-III mine, but later it became a mine of its own, notably through the

[2]This name literally means: Company for Exploitation of Limburg Coal Mines.

efforts of Dutch mining engineer Cornelis Raedts, who later became CEO of Oranje Nassau Mines. Mining of the coal in the ON-IV was, after its closure, completed from the neighbouring ON-III mine (Voncken and De Jong 2003).

1.3.4 Laura and Vereeniging

In the late 19th Century, miller Anton Wackers from Herzogenrath (Germany) and his brother-in-law, Gustav Schümmer, started drilling for coal near the village of Eygelshoven, close to Kerkrade in The Netherlands. Eygelshoven is located just on the Dutch side of the Dutch-German border. Before this time, Count Alfred de Marchant et d'Ánsembourg[3] and his brother Oscar had already made drillings, and had reached Carboniferous rocks at 63 m depth. Wackers and Schümmer, who had bought the land, went on with their search. They found coal at 154 m, and a few weeks later a thick seam a little deeper. They applied for a concession, named "Laura," Laura being the name of the wife of Wackers (and sister of Schümmer). On September 9, 1876 they obtained the rights for the exploitation of an area of 457 ha under the municipalities of Kerkrade, Eygelshoven and Nieuwenhagen. Others had tried also exploration in Eygelshoven (in the so-called Kommerveld[4] area), and were also successful. They named themselves "Vereenigd Gezelschap voor Steenkoolontginning in het Wormdistrict" (i.e. United Party for Coal Exploitation in the Worm District), and they tried to obtain a concession under the name of Vereeniging (i.e. Association, or Union). The Association/Union obtained their concession on February 18, 1877 for an area of 454 ha under the communities Eygelshoven, Übach over Worms and Nieuwenhagen (Voncken and De Jong 2003).

The Worm (German: Wurm) is a small left-bank tributary to the Rur[5] River, and has its springs in the Aachener Wald (i.e. Forest of Aachen), just south of the Germany city of Aachen. It flows through the municipalities of Würselen, Herzogenrath, Übach-Palenberg, Geilenkirchen and Heinsberg before it, to the north of Heinsberg, ends in the Rur (Roer). From Herzogenrath until Übach-Palenberg the Worm River is the boundary between The Netherlands and Germany (Fig. 1.6).

In 1887, the concession Laura was purchased by the Eschweiler Bergwerksverein[6] (EBV) and the holders of Vereeniging. The concessions were joined, and the name "Laura and Vereeniging" (i.e. Laura and Union) was born. In 1899, a company,

[3]De Marchant et d'Ánsembourg is a Dutch noble family, originally of Luxemburgian descent. All members bear the title Count or Countess (Wikipedia 2018a).

[4]A marshy area in the former town of Eygelshoven, now a district of the municipality of Kerkrade.

[5]Often misspelled as *Ruhr.* It is, however, ***not*** named after the German *Ruhr River*, which is an important right-bank tributary of the lower Rhine, joining it at Duisburg-Ruhrort, but the *Rur River* (spelled in Dutch as "*Roer*"). Following recent literature, the spelling "Roer" will be used here as well, in order to avoid confusion. N.B.: The "oe" in "Roer" should be pronounced as the "oo" in for instance "look", or "book".

[6]"Eschweiler Bergwerksverein" means "Eschweiler Mine Association". Eschweiler is a village close to Aachen. See Fig. 1.1.

Fig. 1.6 Map showing the Rur Basin and the course of river Roer and its tributaries, the Wurm (Dutch: Worm), Merzbach and Inde (The Inde is a small river, which is, like the Worm, a tributary of the Roer. The Inde has its springs near Raeren in NE Belgium, and flows through Aachen-Kornelimünster, Eschweiler, and Inden, before ending in the Roer near Jülich). Thin black lines indicate state boundaries. At Roermond (i.e."Roer mouth") the Roer ends in the river Maas (English: Meuse) (*Source* Wikipedia (https://en.wikipedia.org/wiki/Rur#/media/File:Indeverlauf.png), license CC BY-SA 3.0 (https://creativecommons.org/licenses/by-sa/3.0/legalcode), adapted, and partly redrawn)

funded by the Banque d'Outremer[7] from Bruxelles, was founded under Belgian Law, with its seat in Brussels. The company was called "Société des Charbonnages Réunis Laura et Vereeniging S.A.[8]" The head of the company was Albert Thys, a former military man, who had gone into banking. The Banque d Óutremer of Thijs

[7]Itself founded on January 7, 1899.

[8]S.A. means Société Anonyme, which is equivalent to a public limited company (plc). In Dutch, this would be "naamloze vennootschap."

was the greatest shareholder. In 1901 the drilling of the first shaft of the mine Laura started. In 1902 the construction of the second shaft followed.

The shafts were called respectively "Wilhelmina" and "Hendrik," after the Queen of The Netherlands at that time, Wilhelmina, and her prince consort, Prince Hendrik. On old pictures, the names of the shafts are still visible.

After a major renovation, in later years, new headframes were erected (with the hoisting machines in the top), the discharge areas were renewed, and the shafts were—from then onward—indicated by numbers. Shaft 1 was the former Wilhelmina shaft, and shaft 2 the former Hendrik-shaft (Voncken and De Jong 2003).

Through the middle of the Eygelshoven village runs a major fault (the Feldbiss Fault), which not only offsets the coal layers, but also is strongly water-bearing. After approximately 15 years, the construction of a nearby a second mine was started, on the other side of the fault. This was done because in the beginning of the 20th Century, engineers could not safely penetrate the fault zone, because of the water flowing through it (Voncken and de Jong 2003). Only after mining technology had advanced much further, it became possible to construct a drift (tunnel) through the strongly water-bearing Feldbiss Fault, thus connecting the two mines. This second mine was called "Julia", named after the wife of Albert Thijs. The Laura mine started production in 1905, and the Julia mine started production in 1926. The mines were closed in 1968 and 1974 respectively (Voncken and De Jong 2003). Exploitation of the coal seams of the Laura continued after 1968, but from the Julia shafts. For more information about the Feldbiss Fault, and other local tectonic features, see Chap. 4.

1.3.5 Willem and Sophia

The concessions Willem and Sophia were granted in 1860 and 1861 by the government of Dutch Prime Minister Thorbecke to the Nederlandse Bergwerkvereniging ("Dutch Mine Association") in The Hague. The names of the concessions are derived from King Willem III of the Netherlands (1817–1890), and his first wife, Sophia van Württemberg (1818–1877).

The Nederlandse Bergwerkvereniging went bankrupt in 1881 as they could not succeed in constructing a shaft in the difficult and very wet top soil of the concession fields. In the year 1898 the concessions were sold to a Belgian company called Société Anonymes des Charbonnages Néérlandais Willem et Sophia. This included the shafts dug in "De Ham" (see below). These shafts (Ham-I and Ham-II) were incorporated in the Willem-Sophia mine. Using the AGF method (invented by Poetsch) for constructing shafts in wet top soil, this company, following the enormous success of the Honigmann brothers, quickly succeeded in constructing two shafts in the village Spekholzerheide near Kerkrade. In 1902, the mine, named Willem-Sophia, went into production. In 1950, the Melanie concession, situated in Germany, just across the border, was added. The mine produced coal for domestic use, and was one of the smaller Dutch coal mines. The mine had 5 shafts, of which some (Ham-I, Ham-II) were only ventilation shafts. The mine was closed in 1970 (Voncken and De Jong 2003).

1.3.6 The Ham Mine

The Ham mine was a small 19th Century mine near Kerkrade, The Netherlands. The mine existed from 1862 to 1883. The mine is named after the "Ham," a valley through which the rivulet "Anstellerbeek" flows. This rivulet also flows through the Anstel valley, from which it derives its name.

The first mining activities in the concession "Willem," started in 1862 by the Nederlandse Bergwerkvereniging in the hamlet of Ham, were situated close to the community of Kerkrade. The shaft of the mine later vanished, in 1934, under the railway embankment of the so-called "Million Line" railway, which was called like this because of the millions of Dutch guilders it had cost to build this railway in the hilly terrain of South Limburg, necessitating the construction of numerous bridges, viaducts and locally quite elevated railway embankments.

Between 1870 and 1878, several attempts were made to drill a shaft, but efforts were hampered by large amounts of water, which flooded the works regularly. In 1878, a 1 m thick coal layer, dipping 20°, was found at 51 m below ground level. This coal layer was called "Steinknipp" (i.e. "stone cut"). At 61 m depth, a drift was made. In 1881, there was an economic crisis with respect to coal, and the Nederlandse Bergwerkvereninging went bankrupt. As a result, the concessions "Willem" and "Sophia", together with the Ham shaft(s), were offered for sale. As mentioned above, in the year 1898 the concessions were finally sold to a Belgian company called Sociëté Anonymes des Charbonnages Néërlandais Willem et Sophia, which constructed the shafts of the later Willem-Sophia mine.

1.3.7 State-Owned Mines

In 1899 the Dutch government, inspired by the success of the Honigmann brothers, Sarolea and their Oranje Nassau-I (ON-I) coal mine, initiated a commission to investigate if coal exploitation by the Dutch state was desirable. This commission decided in favour of state exploitation. It would make Dutch coal supply less dependent upon foreign countries and/or companies.

After the Dutch parliament agreed with the proposal that was advocated by Minister of Transport, Cornelis Lely, the company "Staatsmijnen in Limburg" (State Mines in Limburg) was founded on the first of May 1902. The first director of the Company was H. J. E. Wenckebach, who had also founded the Dutch steel company Koninklijke Nederlandsche Hoogovens en Staalfabrieken (i.e. "Royal Dutch Blast Furnaces and Steel Factories"), which later became part of the Dutch-British company Corus, and since 2007 has been part of the Indian company Tata Steel.[9]

[9]This steel company was not built in Limburg, where the coal was found, but in IJmuiden, a village that was in the 19th Century constructed at the mouth of the North Sea Canal. The reason for the positioning of the iron and steel works at IJmuiden was that coal for production of cokes, found in The Netherlands, could be easily transported by train to IJmuiden, but the iron ore had to come by

Initially, there were four state mines:

- *Wilhelmina*, (1906) named after the (at that time) Queen of The Netherlands.
- *Emma*, (1911) named after the mother of Wilhelmina, and who had been for a while Queen Regent until her daughter Wilhelmina, (who had become heir to the Dutch throne at the age of 10), had come of age.
- *Hendrik*, (1915) named after the prince consort of Wilhelmina.
- *Maurits*, (1926) named after a famous forefather of the Dutch Royal Family (Maurits van Oranje, 1567–1625, stadtholder and army commander of the Republic of The United Netherlands).

The Wilhelmina Mine started production in 1906. The Emma Mine started production in 1911. The Hendrik Mine started production in 1918, and the Maurits mine started production in 1923. For both the Emma and Hendrik Mines, coal was located at a larger depth than it was in the Wilhelmina mine and the Oranje Nassau mines. On the other hand, contrary to the other privately owned mines, which had produced high rank coal, suited for industrial combustion (power plants) and for domestic heating, the state mines produced coal suited for production of coke and gas, and in 1914, a coke factory for the Emma Mine was opened. Near the Maurits Mine, a second coke factory was constructed. In 1954, the construction of a fifth state mine (named "Beatrix", after the current Crown Princess, and from 1980 to 2013 Queen of the Netherlands) was started near Roermond, but construction was abandoned in 1962. Due to economic reasons, Dutch coal mining ceased in the years 1968–1974 (Messing 1988). The heyday of the Dutch coal mining thus was in the 1940s and 1950s (Fig. 1.7).

The largest Dutch coal mine was the Maurits State Mine, closely followed by the Emma State Mine. The Hendrik Mine was the deepest mine of all Dutch coal mines, and reached a depth of just over 1000 m. The Maurits Mine was for a long time (contrary to the Hendrik and Emma Mines) a mine with two shafts. The third shaft was completed in 1958.

The coal which was mined in South Limburg is largely similar to that mined in the nearby Aachen area, the German Ruhr area, and the Belgian Campine area. As can be seen in Fig. 1.2, the Aachen coal mining area, the South Limburg coal mining area, and the Campine coal mining area were all located in what is in fact one and the same coalfield.

The coal deposits of South Limburg, the Aachen area and the Belgian Campine area are of Upper Carboniferous age, more precisely of Westphalian age (318,30 till 305,0 Ma) (Dinoloket 2017). Chronostratigraphic definitions of the Westphalian differ between sources. Here the definition for the Westphalian by the Dutch internet site Dinoloket[10] is used. The Westphalian is not considered an official stratigraphic

ship from overseas. Transport of the iron ore was much more expensive, due to its higher weight. Also, the positioning of the iron and steel works at IJmuiden was beneficial because of the position by the sea, and the transport possibilities this offered, via the North Sea Canal.

[10]Dinoloket is a website of the Dutch Geological Survey—TNO, where DINO stands for Data en Informatie van de Nederlandse Ondergrond (i.e. Data and Information of the Dutch Subsurface). Loket is Dutch for "counter."

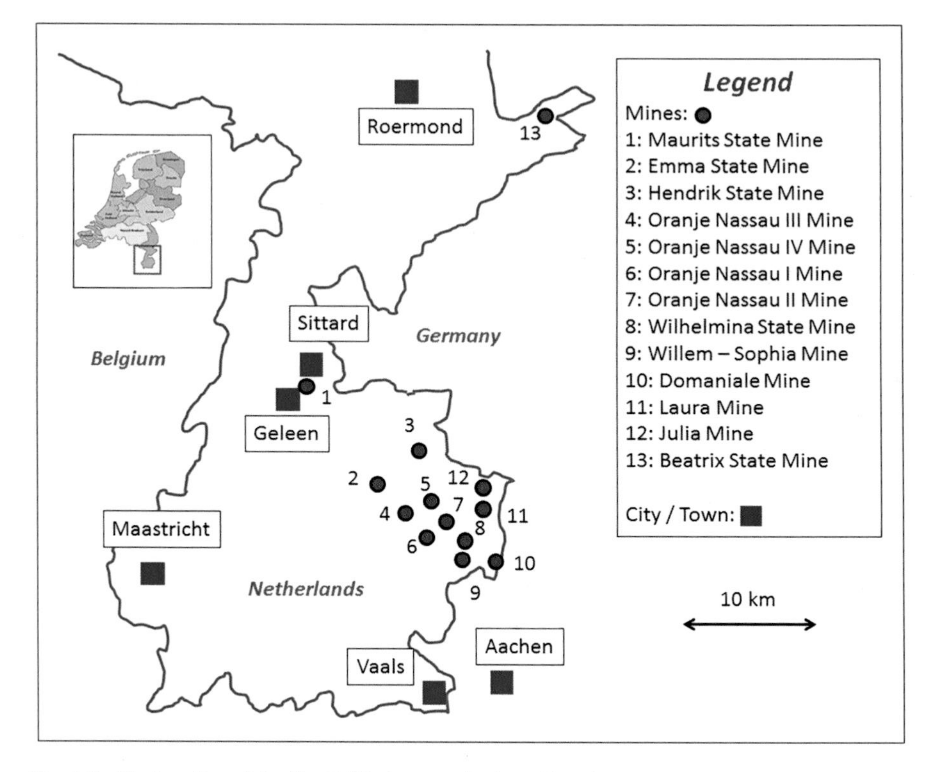

Fig. 1.7 The location of the South Limburg coal mines. Own image

stage any longer by the ICS (International Commission on Stratigraphy), but is used here for convenience, and also to put it into historical context, as it was very much used in the past. See Fig. 1.8.

Ages and correlations of the carboniferous were achieved by use of radiometric dating and conodont[11] stratigraphy (Schmitz and Davydov 2012). Figure 1.8 gives the general stratigraphy of the Carboniferous.

The coal in the Aachen-South-Limburg-Campine coal field has a rank ranging from bituminous coal to anthracite (see also Chap. 2). Table 1.2. lists the different coal types per mine, using local (Dutch) names. "Lean coal" is approximately equivalent to High Rank Coal, also known as semi-anthracite (see Chap. 2). Coking coal is approximately equivalent to what is also called "metallurgical coal," or "fat coal," which in the ICCP[12] Classification is equal to medium rank coal, with 19–28% volatiles, and 87.5–89.5% carbon (see also Chap. 2).

The different types of coal were (per mine):

---

[11]Conodonts (From ancient Greek *kōnos*, "cone", +*odont*, "tooth") are extinct chordates resembling eels, classified in the class Conodonta. They are often used as "index fossils" (Wikipedia 2017a).

[12]ICCP = "International Committee for Coal and Organic Petrology."

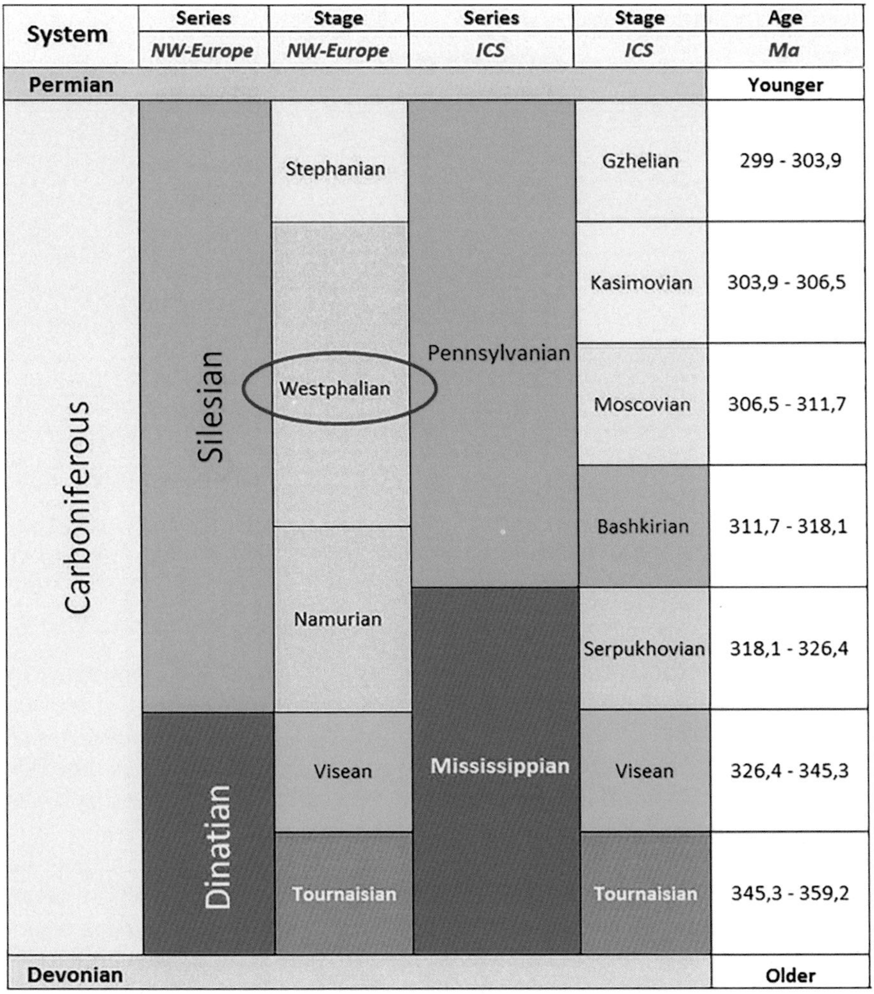

System	Series *NW-Europe*	Stage *NW-Europe*	Series *ICS*	Stage *ICS*	Age *Ma*
Permian					Younger
Carboniferous	Silesian	Stephanian	Pennsylvanian	Gzhelian	299 - 303,9
		Westphalian		Kasimovian	303,9 - 306,5
				Moscovian	306,5 - 311,7
				Bashkirian	311,7 - 318,1
		Namurian	Mississippian	Serpukhovian	318,1 - 326,4
	Dinatian	Visean		Visean	326,4 - 345,3
		Tournaisian		Tournaisian	345,3 - 359,2
Devonian					Older

Fig. 1.8 Stratigraphy of the Carboniferous. ICS is the International Committee on Stratigraphy (http://www.stratigraphy.org). The Westphalian is indicate by the ellipse. [*Source* Wikipedia https://en.wikipedia.org/wiki/Westphalian_(stage) (redrawn)]

1.4 The End of the Dutch Coal Mining Industry

The end of Dutch coal mining came at the end of the 1960s, and the beginning of the 1970s. Due to the discovery of the giant gas field at Slochteren, Groningen in 1959 in The Netherlands, domestic heating became more and more focused on natural gas as a power source. Gas made from degassing coking coal, used for instance for cooking and heating, went into disuse. With lowering of the price of coal on the international market, the exploitation of the Dutch Coal mines became economically unprofitable.

Table 1.2 The Dutch coal mines, and the type of coal mined

Mine	Main type of coal mined	Years active
Domaniale	"Lean coal"	1815–1969
ON-I	"Lean coal"	1899–1974
ON-II	"Lean coal"	1904–1971
ON-III	"Lean coal"	1917–1973
ON-IV	"Lean coal"	1925–1966
Wilhelmina	"Lean coal"	1906–1969
Emma	"Coking coal"	1911–1973
Hendrik	"Coking coal"	1915–1963
Maurits	"Coking coal"	1926–1967
Laura	"Lean coal"	1905–1968
Julia	"Lean coal"	1926–1974
Willem-Sophia	"Lean coal"	1902–1970

The type of coal is placed between quotation marks, as these names are not used as official names any more. More about the naming of coal types and the meaning of the names given above, can be found in Chap. 2

In addition, the coal layers were rather thin, and exploitation was labour-intensive. Miners usually had to crawl along, in the stopes, as they were so low that one could not even sit upright in them (Fig. 1.9). Due to small-scale faulting, mechanized coal mining on a large scale was not possible. It was in 1965, when the at that time Dutch Minister of Economic affairs, J. M. Den Uyl, in a speech given in the City Theatre of Heerlen, announced the closure of all Dutch coal mines.

The mines were successively closed over a period of seven years, ranging from 1967 to 1974. The Hendrik State Mine had already been closed in 1963, but because of an underground connection to the Emma State Mine, coal production went on until 1973. The next mine to close was the Maurits State Mine, located in Geleen—-Lutterade, which came into effect in 1967. The Laura Mine followed in 1968. In 1969 came the closure of the Wilhelmina State Mine, and the Domaniale Mine. The Willem Sophia Mine was closed in 1970. The mine ON-II was closed in 1971, and the ON-III and Emma were closed in 1973. The mine Oranje Nassau IV was itself closed in 1966, but was connected to the mine ON-III, from which production continued until 1973. The Julia Mine closed on December 20, 1974, and the last mine to close was the first that had been opened: the mine Oranje Nassau I, in Heerlen, closed on December 31, 1974.

Fig. 1.9 Miner working in a low stope [*Source* Voncken (Visited May 2018)]

1.5 Coal Mining in the Aachen Region

It is not exactly known when the first coal mining in the Aachen region took place. It is possible that even the Romans mined coal there, to be used for the heating of their bathhouses and villas. There have been several finds of coal made in connection with iron manufacturing places, dating from Roman times, which all point to a winning of coal by the Romans (Wrede and Zeller 1988).

In the Wurm-Inde area (Fig. 1.6), there is evidence for a developed mining industry with draining of groundwater as early as the year 1353. Coal was, in medieval times, only mined there where it outcropped, or where it was found under a thin layer of cover. In the 17th Century, there were many small mines which existed in the Aachen region. Every mine had about three to four people as personnel, and were politically owned by one of the following (Wrede and Zeller 1988):

- The County (Duchy) of Jülich
- Ländchen zur Heyden, owning the quarries near Kohlscheid
- Land Herzogenrath, owning the Voccart mine
- Aachener Reich (i.e. Aachen Realm).

Small mining companies such as these were called "Köhlergesellschaften," meaning "companies of coal miners," "Köhler" is a local, South Limburg/German name for a coal miner. When the near-surface coal layers were exhausted, at the end of the 18th Century, the pumps for water drainage were substantially improved, and mining could then take place up to 100 m depth. Steam engines were first employed at the beginning of the 19th Century. In the 19th Century the demand for coal rose, because of the widely applied steam engines, and the development of railways. The many small coal mines became united into larger companies, such as, for instance, the "Vereinigungsgesellschaft für Steinkohlenbau im Wurmrevier,"[13] founded in 1836, the Eschweiler Bergwerksverein, EBV, founded in 1838, and the Pannesheider Bergwerksverein, founded 1842. The latter company merged in 1907 with the EBV, and continued under that name (Raedts 1974).

Around the end of the 19th Century, compressed air began to be used, and electricity also found its use in coal mining. In the beginning of the 20th Century, coal mining more and more became located in the Worm (German: Wurm) area, and correspondingly lesser in the Inde area. In the beginning and middle of the 19th Century, mines like the Voccart, Anna and Maria were founded. Later the mines Adolf, Carolus Magnus, Carl Friedrich, Carl Alexander and Nordstern were started. Some mines were very small, like the mine Teut (1865–1904). In 1914, the mine Sophia-Jacoba, near Hückelhoven (in the neighbourhood of the Dutch town Roermond[14]) went into production. In 1938 the mine Emil Mayrisch was opened.

The names of these mines were derived as follows:

Adolf	named after Freiherr (i.e. Baron) Adolf von Steffens (1817–1898), from 1871 until 1898 Head of the Eschweiler Bergwerksverein —EBV (Gessen 2018).
Anna	named after the wife of one of the concession holders, cooperating with Eduard Honigmann (Schumacher 2007).
Carl Alexander	named after Carl Röchling (1827–1910), CEO of the steel works Röchling from Völklingen, Germany (Saarland)) and Alexander Dreux (1853–1939), CEO of the Société des Aciéries de Longwy, from Longwy, département de Meurthe-et-Moselle, France, (Bergbaumuseum Grube Anna-2 2018a).
Carl Friedrich	named after engineer Carl Widmann (also mentioned as Weidtmann) and Councilor of Commerce Friedrich Wilhelm Huppertz (Bergbaumuseum Grube Anna-2 2018b).
Carolus Magnus	named after Carolus Magnus, or Charlemagne (747–814) Emperor of the Holy Roman Empire (800–814). Charlemagne often resided in Aachen.
Centrum	named after the concession Eschweiler Centrum (Gessen 2018).
Emil Mayrisch	named after Emil Mayrisch, CEO of the ARBED,[15] a steel company from Luxemburg (Bergbaumuseum Grube Anna-2 2018c).
Gouley	named after "Gute Ley", local name for "good rock" (Eiffelnatur 2018; Bergbaumuseum Anna-2 2018d).
Laurweg	named after the area "Laurweg" in the town Kohlscheid, now part of the municipality of Herzogenrath (Gessen 2018).

[13] Union for Coal Mining in the Wurm Area.

[14] At Roermond, (literally meaning Roer Mouth), the river Rur (Dutch: Roer) ends in the river Maas (Meuse).

[15] ARBED: Aciéries Réunies de Burbach-Eich-Dudelange. The mines in the Aachen area sold coking coal to this Luxemburgian steel company.

Maria	named after Maria Bölling, the wife of Eduard Honigmann - German mining engineer, father of Friedrich and Carl Honigmann, (Schumacher 2007).
Nordstern	German for north star.
Voccart	named after a community of Köhler,[16] which had adopted the name "Fouckert." The name "Fouckert" in time degenerated to Voccart (Raedts 1974).
Reserve	named after the concession name "Eschweiler Reserve" (Gessen 2018).
Sophia Jacoba	named after Mrs. Sophia Schout Velthuijs (1882–1976), the wife of Dr. Frits Fentener van Vlissingen (director of the NEMOS—"*NE*derlandsche *M*aatschappij tot *O*ntginning van de *S*teenkolenvelden,"[17] (Source: Wikipedia 2018b) and Mrs. Jacoba Philippina van Dam (1876–1967). the wife of the director of the mine, Ing. Isaäc Pieter de Vooys, (Source: Stamboom de Vooys 2018).
Teut	named after a pointy piece of land. In local dialect: Teute or Tüte, Tüt. Source: Bergbaumuseum Anna-2 (2018e), and the author's own knowledge of the local dialect.

Below, a table is given with relevant data of the coal mines in the Aachen region (Table 1.3) and a map (Fig. 1.10) indicating the villages were the mines were situated is shown.

Coking coal was partly sold to the Luxemburgian Steelcompany ARBED (Aciéries Réunies de Burbach-Eich-Dudelange, i.e. United Steel Works of Burbach-Eich-Dudelange[18]), which secured a steady market for its purchase. The lean coal and anthracite mined by several other mines was used for domestic heating.

The German mines were damaged during the Second World War, but were able to restart production quite soon. From the beginning of the 1960s, the competition rendered by petroleum and natural gas lead to a downfall in the demand of coal, and this resulted in the closure of many mines. Some mines, however, continued in operation until the 1980s and 1990s: the Anna (closure date 31-12-1983), and the Emil Mayrisch (closure date 18 December 1992).

[16] Köhler: South Limburg/German name for a coal miner in the 17th and 18th Centuries.

[17] i.e. "Dutch Company for the Exploitation of Coal Fields."

[18] Burbach is nowadays a part of the city of Saarbrücken (Germany). Eich is nowadays a suburb of Luxemburg City. Dudelange (Lëtzebuergesch/Luxembourgish: Diddeleng) is a town in the uttermost south of Luxemburg, in the Canton Esch).

Table 1.3 Coal mines in the Aachen region

Mine	Type of coal[a]	Years operative	Town
Centrum	"Lean coal"	13th century–1891	Eschweiler
Gouley	"Lean coal"	1599–1969	Morsbach/Würselen
Laurweg	"Lean coal"	1612–1960	Kohlscheid
Voccart	"Lean coal"	1830–1932	Strass/Herzogenrath
Maria	"Lean coal"	1849–1962	Höngen/Alsdorf
Anna	"Coking coal"	1854–1978	Alsdorf
Reserve	"Coking coal"	1862–1944	Eschweiler/Nothberg
Nordstern	"Coking coal"	1880–1927	Merkstein/Herzogenrath
Teut	"Lean Coal"	1865–1904	Schweilbach (Würselen)
Carl Friedrich	Anthracite	1903–1927	Richterich/Aachen
Adolf	"Lean coal"	1913–1973	Merkstein/Herzogenrath
Sophia Jacoba	Anthracite	1914–1997	Hückelhoven
Carolus Magnus	"Coking coal"	1919–1962	Übach-Palenberg
Carl Alexander	"Coking coal"	1921–1975	Baesweiler
Emil Mayrisch	"Lean coal"	1938–1992	Siersdorf/Aldenhoven

Source Salber (1987)
[a]The type of coal is given between quotation marks, as these names are not the official names of the type of coal any more. More about the naming of types of coal can be found in Chap. 2

1.6 Coal Mining in Belgium

In Belgium, there are 5 coalfields:

- The Campine (Kempen) Coalfield. Mining from 1917–1992
- The Liège Coalfield. Mining from the Middle Ages–1980
- The Charleroi Coalfield. Mining from the Middle Ages–1984
- The Centre Coalfield. Mining from the Middle Ages–1973
- The Borinage Coalfield. Mining from the Middle Ages–1973

These coalfields are shown in Fig. 1.2. Discussed in this work are only the mines located in the Kempen (Campine) region, as the other Belgian coalfields are not adjacent to the South Limburg fields, and are geologically separated from the South-Limburg area by major folds and faults. The Campine coalfield is in the northern part of the Belgian Province of Limburg, and forms—together with the Dutch South Limburg coals, and the German deposits of the Aachen area—one single elongated coalfield (see Fig. 1.2).

As is clear from the above description, Belgian coal mining had already been established for a long time, when around 1830 iron and steel became important (Wikipedia 2017b).

Important for the discovery of the Campine coalfield was a report by Professor Guillaume Lambert, of the University of Louvain. He argued that, by analogy with

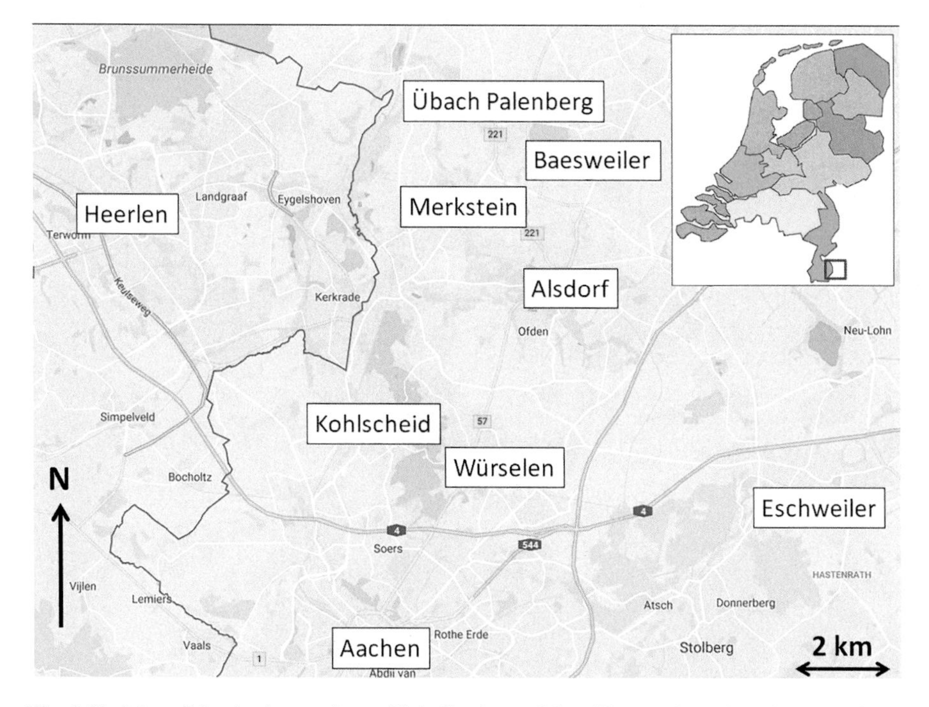

Fig. 1.10 Map of the Aachen region, with indications of the villages where the mines mentioned above were situated. The town of Hückelhoven and the Sophia Jacoba Mine are some 25 km north of Eschweiler, outside the map area. Between Übach Palenberg and Hückelhoven, coal is at a too large depth to mine. This is due to the tectonic disturbance called the Rur Graben (see also Chap. 4) (After Google Maps, adapted)

the German Ruhr area, where a southern, geologically disturbed area was accompanied by a northern, undisturbed area, a northern coal bearing region should exist in Belgium. The southern Liège coalfield might be accompanied by a northern coal field in the Campine area. He therefore advised to continue the drillings for coal from the Dutch border near Geleen in a northwestern direction. This was also due to an interpretation of the results of Dutch drillings near Heerlen. In 1877 a report of a second advisor, André Dumont, appeared, in which Dumont argued that the axis of the just-detected Dutch coal field would bend to the northwest, in the direction of the Campine area. It took a while for Dumont to get further drillings financed, because due to an economic crisis in the 1880s, there was a surplus of coal, and there was little interest in pursuing new exploration. This changed in the 1890s, when the economy turned for the better, and there suddenly appeared to be a shortage of coking coal. In 1898, the S.A. de Recherches et d'Exploitations was established to finance exploration drillings in the Belgian part of Limburg (Minten et al. 1992).

In 1901, The Belgian Campine Coalfields were discovered by André Dumont with a drilling in the municipality of As, northwest of Genk. He found thick layers of coking coal. Two years later 62 drillings had already been carried out. This number,

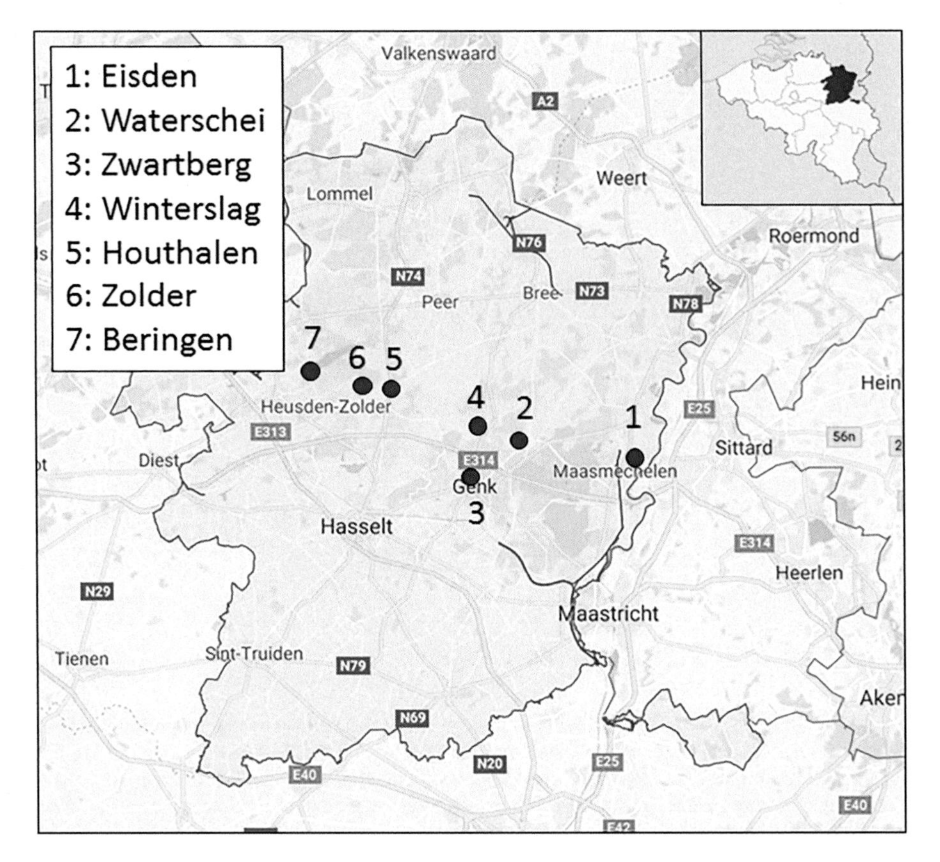

Fig. 1.11 Location of the seven coal mines in the Campine area (After Google Maps and Wikipedia, adapted). The outlined area is the Belgian Province of Limburg

however, had more to do with the fact that, in order to acquire a concession from the authorities, one had to show the potential for a mine, rather than offering an argument based purely on scientific reasoning. As a result, in 1906 ten concessions were granted (Minten et al. 1992):

1. Concession André Dumont, 2950 ha
2. Concession Les Liégois, 4180 ha
3. Concession Helchteren, 3240 ha
4. Concession Zolder, 3820 ha
5. Concession Genk-Sutendael, 3800 ha
6. Concession Beeringen-Coursel, 4950 ha
7. Concession Sainte Barbe, 2170 ha
8. Concession Guillaume Lambert, 2740 ha
9. Concession Houthalen, 3250 ha
10. Campine State concession, 49,172 ha.

Table 1.4 Coal mines in the Campine area

Mine	Type of coal[a]	Years active
Winterslag	"Coking coal"	1917–1988
Beringen	"Coking coal"	1922–1989
Eisden (Limburg-Maas mine)	"Coking coal"	1923–1987
Waterschei (André Dumont mine)	"Coking coal"	1924–1987
Zwartberg (Les Liégeois mine)	"Coking coal"	1925–1966
Zolder	"Coking coal"	1930–1992
Houthalen	"Coking coal"	1938–1992

Source Minten et al. (1992)
[a]The type of coal is given between quotation marks, as these names are not the official names of the type of coal any more. More about the naming of types of coal can be found in Chap. 2

Half a year later, as a result of cooperation between financiers, concession holders, and industrials. already six companies were established, which would go on to exploit 8 concessions (Minten et al. 1992).

In the years 1917–1932, seven mines arose in this area (all named after the municipality in which they were located): Winterslag (1917), Beringen (1922), Eisden (1923), Waterschei (1924), Zwartberg (1925), Zolder (1930), and Houthalen (1938). Data are from Minten et al. (1992). Figure 1.11 shows a map on which the locations of the mines are indicated (Table 1.4).

The mines have, after closure, largely been demolished, but, in contrast to The Netherlands, the headframes at the shafts have been left standing. One mine has been left completely intact, and serves as a museum: the Mine Museum at Beringen (http://www.mijnmuseum.be/).

References

Agricola G (1556) De Re Metallica. Image from: http://www.vmine.net/mining-heritage.com/agricola/book5.htm. Visited May 2018

Bergbaumuseum Grube Anna-2 (2018a) Grube Carl-Alexander. http://www.bergbaumuseum-grube-anna2.de/index.php/grube-carl-alexander/. Visited May 2018 (In German language)

Bergbaumuseum Grube Anna-2 (2018b) Grube Carl-Friedrich. http://www.bergbaumuseum-grube-anna2.de/index.php/grube-carl-friedrich/. Visited May 2018 (In German language)

Bergbaumuseum Grube Anna-2 (2018c) Grube Emil Mayrisch. http://www.bergbaumuseum-grube-anna2.de/index.php/grube-emil-mayrisch/. Visited May 2018 (In German language)

Bergbaumuseum Gruba Anna-2 (2018d) Grube Gouley. http://www.bergbaumuseum-grube-anna2.de/index.php/grube-gouley/. Visited May 2018. (In German language).

Bergbaumuseum Gruba Anna-2 (2018e) Grube Teut. http://www.bergbaumuseum-grube-anna2.de/index.php/grube-teut. Visited May 2018. (In German language).

Dinoloket (2017) Chronostratigraphy chart. https://www.dinoloket.nl/carboniferous. Visited December 2017.

Eiffelnatur (2018) Steenkolenmijn Gouley. http://www.eifelnatur.de/Niederl%E4ndisch/Seiten/Steenkolenmijn%20Gouley.html In Dutch Language. Visited May 2018. (In Dutch language).
Geo-engineer (2017) Artificial Ground Freezing. http://www.geoengineer.org/education/web-based-class-projects/select-topics-in-ground-improvement/ground-freezing?start=1. Visited Dec 2017
Gessen R (2018) Eisenbahnen der Region Aachen-Düren-Heinsberg. http://www.gessen.de. Visited May 2018 (In German language)
http://www.eifelnatur.de/Niederl%E4ndisch/Seiten/Steenkolenmijn%20Gouley.html. Visited May 2018 (In Dutch language)
Patijn RJH, Kimpe WFM (1961) Explanation to the map of the Carboniferous surface, the sections and the map of the overburden of the South-Limburg mining District and State Mine Beatrix with the surrounding area. Mededelingen van de geologische stichting; Serie C, 1-1-No. 4
Messing FAM (1988) Geschiedenis van de mijnsluiting in Limburg. Noodzaak en lotgevallen van een regionale herstructurering, 1955–1975. Martinus Nijhoff Publishers, Leiden, 707 pp (In Dutch language)
Minten L, Raskin L, Soete A, Van Doorslaer B, en Verhees F (1992) Een eeuw steenkool in Limburg. Uitgeverij Lannoo, Tielt, 280 pp (In Dutch language)
Poetsch FH (1884) Method and apparatus for sinking shafts through quicksand. No. 300891-Patented 24 June 1884. http://www.google.com/patents/US300891?hl=nl
Raedts CEPM (1974) De opkomst, de ontwikkeling en de neergang van de steenkolenmijnbouw in Limburg, Maaslandse Monografieën, Van Gorcum BV, Assen, 203 pp (In Dutch Language)
Salber D (1987) Das Aachener Revier—150 Jahre Steinkohlenbergbau an Wurm und Inde. Verlag Schweers and Wall, Aachen, 128 pp
Schmitz MD, Davydov VI (2012) Quantitative radiometric and biostratigraphic calibration of the Pennsylvanian-Early Permian (Cisuralian) time scale and pan-Euramerican chronostratigraphic correlation. Geol Soc Am Bull 124(3–4):549–577
Schumacher M (2007) Alsdorf und Anna, In: Glückauf, Nr. 26, ISSN 1864-5526, pp 4–20 Bergbaumuseum Grube Anna e.V. (Publisher): Anna Glückauf Berichte-Mitteilungen-Nachrichten (In German language)
Stamboom de Vooys (2018) https://www.genealogieonline.nl/stamboom-de-vooys/I26.php. Visited May 2018 (In Dutch language)
Stichting “De Mijnen” (2017) https://www.demijnstreek.net/index.php. Visited Dec 2017
Voncken JHL (2003) Ontstaansgeschiedenis van de Steenkolenwinning in Limburg. Nat Resour 10(4):34–37 (In Dutch language)
Voncken JHL, De Jong TPR (2003) Steenkolenwinning in Nederland—Coal Mining in The Netherlands. TU Delft Repository. https://repository.tudelft.nl/islandora/object/uuid:df6d0d73-46b8-46e1-8da2-1e644bd60c18/datastream/OBJ/download
Wikipedia (2017a) Conodonts. https://en.wikipedia.org/wiki/Conodont. Visited Aug 2017
Wikipedia (2017b) History of coal mining/Belgium. https://en.wikipedia.org/wiki/History_of_coal_mining#Belgium. Visited Dec 2017
Wikipedia (2018a) De Marchant et d’Ansembourg. https://nl.wikipedia.org/wiki/De_Marchant_et_d%27Ansembourg#Nederlandse_tak. Visited Oct 2018 (In Dutch language)
Wikipedia (2018b) Frits Fentener van Vlissingen. https://nl.wikipedia.org/wiki/Frits_Fentener_van_Vlissingen_(1882. Visited May 2018 (In Dutch language)
Wrede V, Zeller M (1988) Geologie der Aachener Steinkohlenlagerstätte (Wurm- und Inderevier). Geologisches Landesamt Nordrhein-Westfalen, 77 pp (In German language)

Chapter 2
The Origin and Classification of Coal

Abstract This chapter describes the process of coalification, which gradually turns plant debris into coal, involving heat, pressure and the effects of time. Chemical changes during peatification and coalification are described, and also structural changes in coal during coalification are covered (cleats and their development). The environments in which coal can be formed are described in detail. The chapter gives an overview of the classification of coal, using several systems from the past, as well as the present-day, official one. Methods by which coal can be studied and classified are described, including coal petrography, vitrinite reflection measurement, and chemical and physical characterization methods (ultimate and proximate analyses). The concept of macerals is also covered.

Keywords Coal forming environments · Coalification · Classification of coal · Cleats · Chemical and physical characterization · Macerals

2.1 Introduction

Coal is classified as a biogenic sedimentary rock within the group of sedimentary hydrocarbons. It is a combustible black rock consisting mainly of carbon.

Coal is formed from the remains of plants, by a process called *coalification*. The whole process starts with the remains of dead plants, which must be buried in an oxygen-poor or oxygen-free environment, to avoid complete decomposition. Usually, these are swamp-type environments. The coalification process takes place over millions of years.

In nature, coal is present in geological formations varying in age from Carboniferous till Miocene (358.9 Ma until 5.33 Ma). There have been two major periods in the Earth's history, when large coal deposits were formed throughout almost the whole world: the Upper-Carboniferous—Permian period and the Tertiary (Paleogene) period. See Fig. 2.1.

For the classification of coals, and explanation of the term brown coal/lignite, see Fig. 2.4. Coals from the Upper Carboniferous and Permian are so-called *hard coals*. These coals are not easily deformable. Coals from the Tertiary (Paleogene and

J. H. L. Voncken, *Geology of Coal Deposits of South Limburg, The Netherlands*, SpringerBriefs in Earth Sciences, https://doi.org/10.1007/978-3-030-18286-1_2

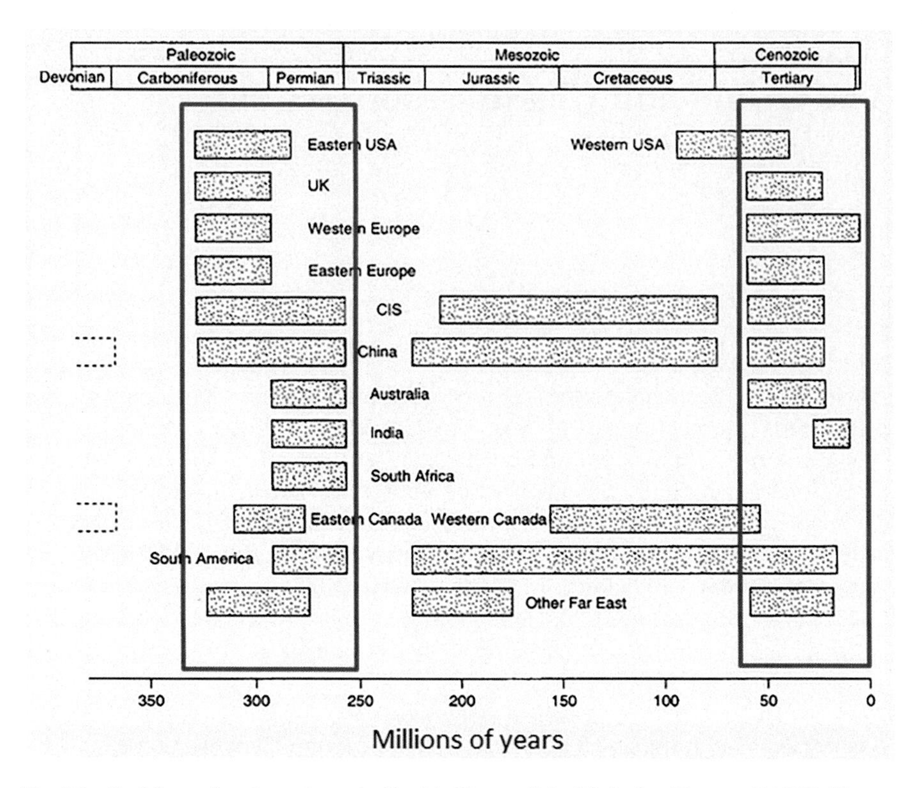

Fig. 2.1 Coal formation throughout the Earth's History [Modified after Thomas (2002)]. The two large rectangles indicate the periods when there was major coal formation throughout almost the whole world

Neogene) are mainly so-called *brown coals* (or *lignite*). Such coals are softer, friable materials. Coals from the Mesozoic are found largely in the USA, CIS,[1] China, South America, and other countries in the Far East (Thomas 2002). They are usually *hard coals*, although *brown coal (lignite)* also occurs. Coals from the Tertiary (Paleogene and Neogene) are mainly *brown coals.*

As a matter of fact, formation of a coal seam is an unusual geological occurrence (King 2016). For this to happen, one of the following conditions must apply:

1. A rising water level that keeps perfect pace with the rate of plant debris accumulation.
2. A subsiding landscape that keeps perfect pace with the rate of plant debris accumulation.

Most coal fields have probably been formed under condition 2, in a delta environment.

[1]CIS = Commonwealth of Independent States (former Soviet Union).

The coincidence of factors which are necessary for coal formation is so rare that it is understandable that major coal formation has taken place in only a very limited number of periods in the history of the Earth (King 2016). See also Fig. 2.1.

In Northwestern Europe, productive coal seams are mostly of Upper-Carboniferous (Pennsylvanian) age, ca. 300 Ma old. Brown coals or lignites are usually present in formations younger than the Cretaceous period (<ca. 65 Ma).

There was dominant coal formation during the Carboniferous. This period of geological history is therefore named Carboniferous after the ubiquitous coal deposits found world-wide. There are several reasons for this dominant coal formation during the Carboniferous:

- The climate of the areas that later produced coal, was warm and humid (tropical), and therefore favorable for plant growth in stagnant swamps.
- Swamps are usually low in oxygen, and thus reduced decomposition creates carbon accumulations.
- Because of the rise of the water table and/or subsidence, swamps were submerged at regular intervals. This lead to the formation of a cover of sand, clay and other debris which protected the organic layer from quick erosion.

The quality of each coal deposit is determined by *temperature and pressure* and by the *length of time in formation*, which is referred to as its "*organic maturity*."

The process of increase in organic maturity is also known as *Coalification.*

2.2 Formation of Peat

First, peat is formed. This is the source for coal. The formation of peat depends on permanently present stagnant groundwater, above or close to the ground surface, so that accumulated plant material will not decompose. This occurs generally in coastal flatlands where sea water dams up fresh water coming off of the land. Such areas are called swamps. They are associated with sea coasts or with the shores of large inland lakes.

Fundamental for the growth of thick peat deposits is a slow continuous rise of the groundwater table (Thomas 2002), where the following conditions are present:

1. The groundwater table should maintain a more or less constant relationship between the water table and the upper surface of the peat deposit.
2. Protection of the mire (by beaches, sand bars and the like) against major and prolonged flooding by the sea, and by natural levees against river flood-waters.
3. Physiographic and other conditions affecting the supply of sediments, which allow peat to form over prolonged periods, without interruption by the deposition of fluviatile sediments.

2.3 Coal-Producing Environments

There are several different groups of coal-producing environments. These are the braid plain, the alluvial valley, the upper and lower delta plain, the barrier beach/strand plain system, and the estuary.

There are two types of peat-producing wetlands. The first type is ombrogenous peatlands, or mires. They owe their origin to rainfall. The second type is topogenous peatlands. They owe their origin to a place and its surface or groundwater regime (Diessel 1992).

Water-logging of vegetation is caused by ground water. Ombrogenous sites are of greater extent but are less varied in character. The inorganic content, i.e. sediments, of mires increases in the topogenous rheotrophic (flow-fed) mires (Diessel 1992).

Characteristics of coals are influenced during peat formation by the following factors (see e.g. Orem and Finkelman 2005):

1. type of deposition,
2. peat-forming plant communities,
3. nutrient supply,
4. acidity,
5. bacterial activity,
6. temperature,
7. redox potential.

Preservation of organic matter is influenced by the following factors (see e.g. Orem and Finkelman 2005):

1. Rapid burial,
2. Anoxic conditions in the water-logged section of the peat profile,
3. Higher temperatures giving higher decay, occurring in warm climates,
4. Humification rates, which are dependent upon the acidity of the ground water,
5. High acidity suppressing microbial activity in the peat.

An organic-rich system will become anoxic faster than an organic-poor system will, due to oxygen consumption (CO_2-formation). Furthermore, anaerobic metabolic processes (fermentation, methanogenesis, sulfate, nitrate, iron and manganese reduction) are in essence less efficient in breaking down organic matter than aerobic processes. Therefore, a greater part of the organic matter will survive destruction when anaerobic conditions prevail (Orem and Finkelman 2005).

2.4 Coalification

Coalification is the process by which peat is transformed into coal. The process of transforming vegetable matter into coal usually occurs in two main steps: the biochemical and the physicochemical stage of coalification (Stach et al. 1982; Diessel

1992). In the biochemical stage, organisms initiate and assist in the chemical decomposition of vegetal matter, and in its conversion into peat and brown coal. Here, the type of plant material and the environmental conditions influence the type of coal formed. Different biological, chemical and physical constraints result in different peat types, which—during the following physicochemical coalification—are transformed into different types of coal (Diessel 1992).

Coalification is in fact the metamorphosis of coal, as a result of burial (increase of pressure and increase in temperature, and influx of thermal water or brines). The most important controlling parameters are temperature and time. Pressure resulting from the sediment overburden is generally considered to be important only because it serves as a means to control temperature (Orem and Finkelman 2005).

The quality of coal is determined by temperature and pressure, and by the length of time in formation, which is referred to as its "organic maturity." The organic maturity is referred to as the rank of coal.

Coal rank can be determined by chemical parameters, such as amount of organic carbon (dry, ash-free), the atomic H/C ratio, and the atomic O/C ration (Orem and Finkelman 2005). Coal rank can also be determined by measuring its so-called vitrinite reflection (see Sect. 2.5).

In general, the process of coalification is considered to have five main stages resulting in the following types of coal (from low rank to high rank) (Table 2.1).

Bituminous coal contains bitumen.[2] Anthracite (the name derives from the Greek anthrakítēs (ἀνθρακίτης), meaning "coal-like") is nearly completely carbon.

In several Western European countries, as well as in the USA, other names for coal were in use, and in both the USA and Europe a detailed classification was created. The table below gives an overview of the different names for different types of coal. As can be seen, the categories are based on the amount of volatiles in the coal, and its carbon content. The ICCP stands for "*International Committee for Coal and Organic Petrology*" (Table 2.2).

Ascending from low to high rank, the amount of volatiles in coal decreases, and consequently the carbon percentage increases. See Table 2.3 for a detailed description of the coalification processes per stage. Fat coal is also known as metallurgical coal, and can be used to make coke (which is the reducing agent used, for instance, in the blast furnace process for iron making). Low-rank coals, such as lignite and subbituminous coals, are typically softer, friable materials with a dull, earthy appearance. Higher-rank coals are generally harder and stronger. Anthracite is at the top of the rank scale, and has a vitreous luster (see Table 2.1). Rank classes can also be given in terms of vitrinite[3] reflectance.

High volatile bituminous coal is somewhat comparable to low-rank coal or flame coal (Table 2.1). Figure 2.2 shows coal rank versus vitrinite reflectance.

---

[2]Bitumen is a black viscous mixture of hydrocarbons obtained naturally or as a residue from petroleum distillation.

[3]Vitrinite is a so-called maceral group. Macerals are microscopically recognizable constituents of coal, analogous to minerals in a rock. See Sect. 2.5 for detailed information.

Table 2.1 The different stages of coalification and image of the coal per stage

Rank	Image
Peat: accumulated organic material, possibly mixed with some sand. This is the starting material (Image from Wikimedia Commons. Photo by Jeff de Longe)	
Brown coal or lignite: low organic maturity and dark black to various shades of brown (Image from Stephen Hui Geological Museum, Department of Earth Sciences, The University of Hong Kong. Used with permission)	
Sub-bituminous coal: progressive increase in organic maturity and removal of water. (Image from nl.depositphotos.com. Used with permission.)	
Bituminous or hard coals: further chemical and physical change, gas is slowly removed. The coal becomes harder and blacker (Image from nl.depositphotos.com. Used with permission)	
Anthracite: Progressive increase in the organic maturity, all gases and many C–O–H chains are destroyed. The coal has become quite shiny (Image from nl.depositphotos.com. Used with permission)	

See Table 2.3 for a detailed description of the coalification processes per stage

Table 2.2 Overview of the different coal names and classifications

German classification	Dutch classification	English classification	USA classification	ICCP classification	Volatiles %	Carbon %
Torf	Turf	Peat	Peat	Peat	>60	60
Braunkohle	Bruinkool	Lignite	Lignite	Low rank coal	45–60	60–75
Flammkohle	Vlamkool	Flame coal	Sub-bituminous coal/Medium volatile bituminous coal	Low rank coal	40–45	75–82
Gasflammkohle	Gasvlamkool	Gas flame coal	Medium volatile bituminous coal	Medium rank coal	35–40	82–85
Gaskohle	Gaskool	Gas coal	Medium volatile bituminous coal	Medium rank coal	28–35	85–87.5
Fettkohle	Vetkool	Fat coal	Medium volatile bituminous coal	Medium rank coal	19–28	87.5–89.5
Esskohle	Esskool	Forge coal	Low volatile bituminous coal	Medium rank coal	14–19	89.5–90.5
Magerkohle	Magerkool	Non baking coal/lean coal	Semi-Anthracite	High rank coal	10–14	90.5–91.5
Anthrazit	Anthraciet	Anthracite	Anthracite	High rank coal	<10	91.5–98

Table 2.3 lists the predominant physical processes and predominant physico-chemical changes of coal during the coalification stages.

The coals which were mined in South Limburg were partly semi-anthracite (mines ON-I to ON-IV, Laura, Julia, Willem-Sophia, Domaniale), and partly medium-volatile bituminous coal or fat coal (State Mines, with the exeption the Wilhelmina mine, which produced coal for domestic heating). Coals mined in the Aachen Region varied from medium rank coal (fat coal) to anthracite. Coals mined in the Campine area were all medium-rank coal (fat coal).

2.5 Coal Petrography

Coal can be microscopically studied, in order to classify the coal, and also to investigate the presence and nature of impurities, to determine its rank, and to determine which type of coal it is. This kind of microscopic study is called coal petrography.

Table 2.3 Major stages of development from peat to meta-anthracite (Thomas 2002)

Coalification stage	Approximate ASTM rank range	Predominant processes	Predominant physico-chemical changes
Peatification	Peat	Maceration Humification Gelification Fermentation Concentration of resistant substances	Formation of humic substances Increase in aromaticity (Wikipedia 2017)
Dehydration	Lignite to subbituminous	Dehydration Compaction loss of O-bearing groups Expulsion of—COOH, CO_2, and H_2O	Decreased moisture contents and O/C ratio Increased heating value Cleat growth
Bituminisation	Upper subbituminous A to high volatile bituminous A	Generation and entrapment of hydrocarbons depolymerization of matrix Increased hydrogen bonding	Increased vitrinite R_0 Increased fluorescence Increased extract yields Decrease in density and sorbate accessibility Increased strength
Debituminisation	Uppermost high volatile A to low volatile bituminous	Cracking Expulsion of low molecular weight hydrocarbons, especially methane	Decreased fluorescence Decreased molecular weight of extract Decreased H/C ratio Decreased strength Cleat growth
Graphitisation	Semi-anthracite to anthracite to meta-anthracite	Coalescence and ordering of pre-graphitic aromatic lamellae Loss of hydrogen Loss of nitrogen	Decrease in H/C ratio Stronger XRD-peaks Increased sorbate accessibility Anisotropy Strength ring condensation Cleat healing

Coal petrography[4] (or coal petrology[5]) is the study of the organic and inorganic constituents of coal, and their transformation as a result of metamorphism. Coal petrography finds its use in the study of the depositional environments of coal, geological studies of coal, and the investigation of coals for industrial utilization (for instance, coke production).

[4]Petrography: (graphical) description of rock.

[5]Petrology: knowledge or study of rocks.

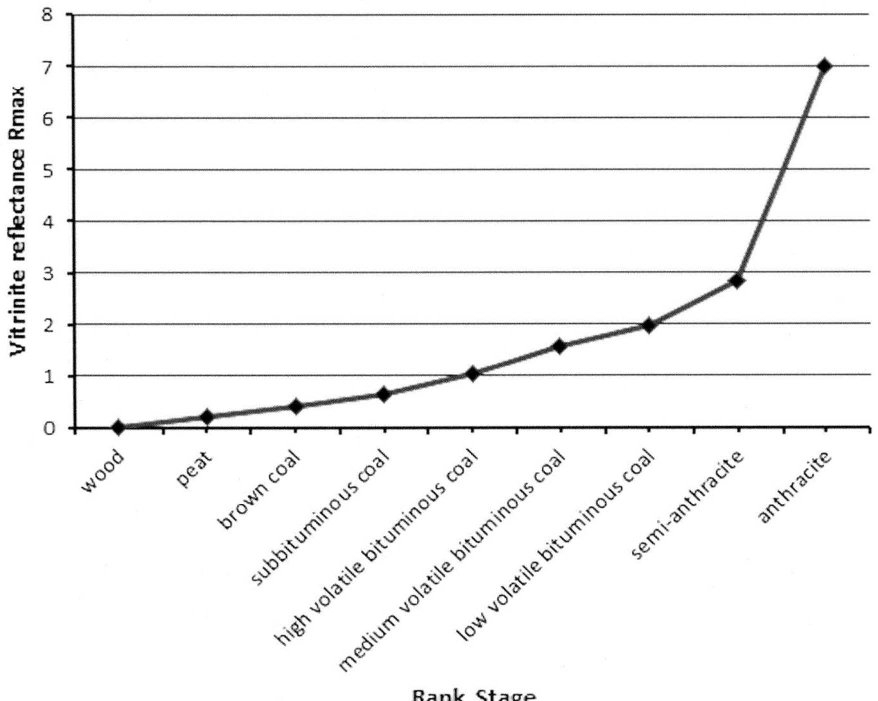

Fig. 2.2 Vitrinite reflectance versus different material: wood, peat and different coal ranks (Data are from Thomas 2002)

In coal, many constituting compounds can be recognized, which are similar to minerals in a rock. These constituents are called *macerals*. The *definition of a maceral* is given below:

> …macerals are organic substances, or optically homogenous aggregates of organic substances, possessing distinctive physical and chemical properties, and occurring naturally in the sedimentary, metamorphic, and igneous materials of the Earth…. (Spackman 1958)

The term maceral[6] was first coined by Stopes[7] (1935). A formal classification of the macerals and the definition of their individual names were made at a congress in the Dutch town of Heerlen, in 1935. Heerlen was the central town of Dutch coal mining, where also the headquarters of Dutch State Mines (DSM)[8] were (and still are) located. This classification system, which is still in use, is referred to as the

[6]Maceral is derived from Latin *Macerare*, i.e. to macerate, to separate.

[7]Marie Carmichael Stopes (15 October 1880–2 October 1958). British author and palaeobotanist.

[8]Dutch State Mines (DSM) were originally called "Nederlandse Staatsmijnen," and in their beginning years, "Staatsmijnen in Limburg" (State Mines in Limburg).

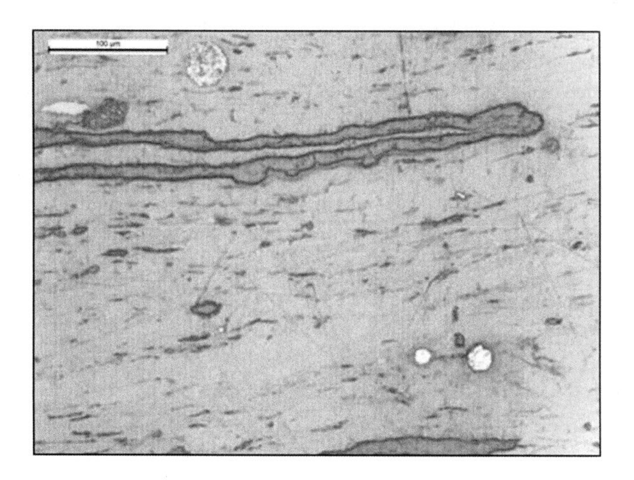

Fig. 2.3 Coal Petrography: a microscopic image of bituminous coal. Sample COP10 (Coal Petrography Collection of Delft University of Technology, Department of Geoscience and Engineering). Own image. Scale bar is 100 μm. See text for explanation

Stopes-Heerlen Classification of Macerals. By convention, maceral names always have an "*-inite*" suffix.

Examples are *telinite, sporinite* and *micrinite.* The type of coal is dependent upon the amount and category of macerals.

Coal can be divided into three groups:

1. Huminite/Vitrinite—woody material
2. Exinite (Liptinite)—spores, resins and cuticles
3. Inertinite—oxidized plant material

The name *kerogen* is used for the same elements in a broader sense, and particularly for dispersed organic material.

The rank of coal can also be determined by the amount of reflectance of the vitrinite compound (vitrinite reflectance). This can also be perceived from the pictures of the 5 main stages of coalification (Table 2.1). Peat is not shiny at all. Anthracite, however, is clearly a shiny material. Vitrinite reflection (Fig. 2.2) is measured microscopically, on a polished section of coal, using a reflective light microscope equipped with a photo-electric cell for measurement, and an indicating device.

In Fig. 2.3, a microscopic image of a bituminous coal is shown for illustration. The coal sample, from Leeds, England, is housed in the Coal Petrography Collection of Delft University of Technology. Visible here is vitrinite (mainly collinite, see Table 2.4) with a megaspore (dark, loop-shaped material in the upper part of image, and smaller parts, left from the center, and in the bottom of the image). Three spots with framboidal[9] pyrite (FeS_2) (bright circularly shaped material) are also visible.

[9]The term *framboid* is derived from the French *la framboise*, Dutch: *framboos*, German: *Himbeere*, meaning 'raspberry', due to the appearance of the structure under magnification.

Table 2.4 The Stopes-Heerlen classification of macerals of hard coals (from Thomas 2002)

Maceral group	Maceral	Submaceral
		Telocollinite
Huminite/Vitrinite	Telinite	Gelocollinite
	Collinite	Desmocollinite
		Corpocollinite
	Sporinite	
	Cutinite	
	Suberinite	
	Resinite	
Exinite (Liptinite)	Alginite	
	Liptodetrinite	
	Fluorinite	
	Bituminite	
	Exudatinite	
	Fusinite	
	Semifusinite	
Inertinite	Macrinite	
	Micrinite	
	Sclerotinite	
	Inertodetrinite	

2.6 Chemical Changes During Peatification and Coalification

2.6.1 Peatification

Peatification, also called "biochemical coalification," includes microbial and chemical alteration. In the peatigenic layer (which is the peat layer from the surface of the peat to a depth of approximately 0.5 m.), the most severe alteration takes place. In this layer, aerobic bacteria, actinomyces (a group of anaerobic bacteria) and fungi are active. With increasing depth, these organisms disappear, and are replaced by only anaerobic bacteria. As the easily assimilated substances gradually disappear, with increasing depth, microbial life is also reduced, and finally completely disappears. The latter is usually the case at depths of less than 10 m. At larger depths, only chemical changes occur, which consist mainly of condensation, polymerization and reduction (Zeng 2016). Humification, a process forming the humic substances,[10] is the most important process during peatification. In a peat profile, the carbon content rapidly increases with depth. This is caused by the microbiological decomposition of oxygen-rich materials within this layer, particularly cellulose and hemicellulose. This results in an enrichment of relatively carbon-rich lignin and in the newly-formed humic acids. A rise in carbon content from 35–50% to 55–60% is quite possible in

[10]Humic substances are major components of natural organic matter. They make up much of the characteristic brown color of decaying plant debris, and contribute to the brown or black color in surface soils.

the peatigenic layer. At greater depth, however, the carbon content hardly changes (Zeng 2016). Also, with increasing depth, the moisture content goes down rapidly. Free cellulose can be found in peats, and is an indicator for the degree of diagenesis of peat. As a result, moisture content, carbon content and the presence of free cellulose are parameters used to discriminate between peat and soft brown coals (Zeng 2016).

2.6.2 *Coalification*

The coalification behaviour is not the same for every maceral. This can be shown in the so-called Van Krevelen[11] Diagrams (van Krevelen 1950). Bustin et al. (1983) adapted the basic diagram (Fig. 2.4).

Carbon dioxide, water and some methane are yielded during coalification. If coals are rich in resinite, some liquid hydrocarbons are also produced during this stage (Zeng 2016). Petrographic changes at the boundary of the dull/bright brown coal stage are very pronounced. Typical chemical changes are gelification (vitrinization) of humic substances. The coals, until this stage having a brown color, become black and lustrous. In the later bituminous coal stages, the amount of volatile matter decreases rapidly, due to the removal of aliphatic[12] and alicyclic[13] groups, and to increasing aromatization[14] of humic complexes (Zeng 2016) (Fig. 2.4).

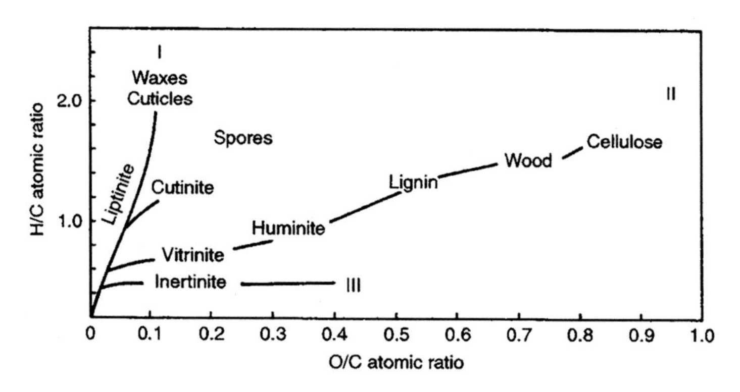

Fig. 2.4 Coalification tracks of different macerals based on H/C versus O/C atomic ratio [a so-called Modified Van Krevelen Diagram] (From Bustin et al. 1983)

---

[11] Dirk Willem van Krevelen (1914–2001) was a prominent Dutch chemical engineer, and a coal and polymer scientist.

[12] Aliphatic groups: non-aromatic groups.

[13] Alicyclic groups: groups that are both aliphatic and cyclic.

[14] Aromatization: the process where an aromatic compound is formed. An aromatic compound is a cyclic (ring-shaped), planar (flat) molecule with a ring of resonance bonds (bonds with a free electron). An example of an aromatic compound is benzene.

Finally, the anthracite stage is characterized by a rapid decrease of the hydrogen content, and thus of the atomic H/C ratio, and correspondingly by a strong increase of reflectivity (see Fig. 2.2) and of optical anisotropy (Zeng 2016).

During peatification and coalification, several structural groups in coal diminish in quantity and finally disappear. Hydroxyl is strongly present in peat, but diminishes in higher rank coals, until it has vanished completely in anthracite. All other groups show an increase during the early stages of coalification, but a decline followed by disappearance in later stages (Chen et al. 2012).

Aromatic CH_x starts to decline at the boundary between medium and low-volatile bituminous coals. The presence of aliphatic CH_x is most prominent in high-volatile bituminous A coals ($R_o \approx 1\%$). $C = C$ and oxygenated groups are most abundant in subbituminous coals ($R = 0.43\%$). Aromatic $C = C$ and CH_x are still present in anthracite (Chen et al. 2012).

Oxygenated groups increase at first, but strongly decrease at the stages of subbituminous and high volatile bituminous coals (Petersen et al. 2008). The initial increase is due to enhanced abundance of carboxyl,[15] ketone,[16] and ether[17] groups. The later decrease is most likely due to decarboxylation and to dehydroxylation (Chen et al. 2012).

2.7 Structural Changes in Coal During Coalification

One of the most notable aspects of coal layers is the presence of sets of fracture surfaces, known as “cleats”. Cleats are fractures that usually occur in two sets, which are, in most instances, mutually perpendicular and are also perpendicular to bedding. These are face cleats and butt cleats. Face cleats are through-going fractures, and are formed first. Cleats that end at intersections with through-going cleats are formed later, and are termed butt cleats. These fracture sets and also partings along bedding planes lead to coal which has a blocky character. The following conventional terms are used for cleats: length is dimension parallel to cleat surface and parallel to bedding; height is parallel to cleat surface and perpendicular to bedding. Aperture is the dimension perpendicular to fracture surface. Spacing between two cleats (of the same set) is the distance between them at right angles (Laubach et al. 1998).

These different features are illustrated in Fig. 2.5a, b and c.

[15] A carboxyl groups is a (C(=O)OH)-group.

[16] A ketone (alkanone) is an organic compound with the structure RC(=O)R′, where R and R′ can consist of a variety of carbon-containing substituents.

[17] An ether is an organic compounds that contains an ether group: an atom connected to two alkyl or aryl groups. An alkyl substituent is an alkane missing one hydrogen. An aryl is any functional group or substituent derived from an aromatic ring.

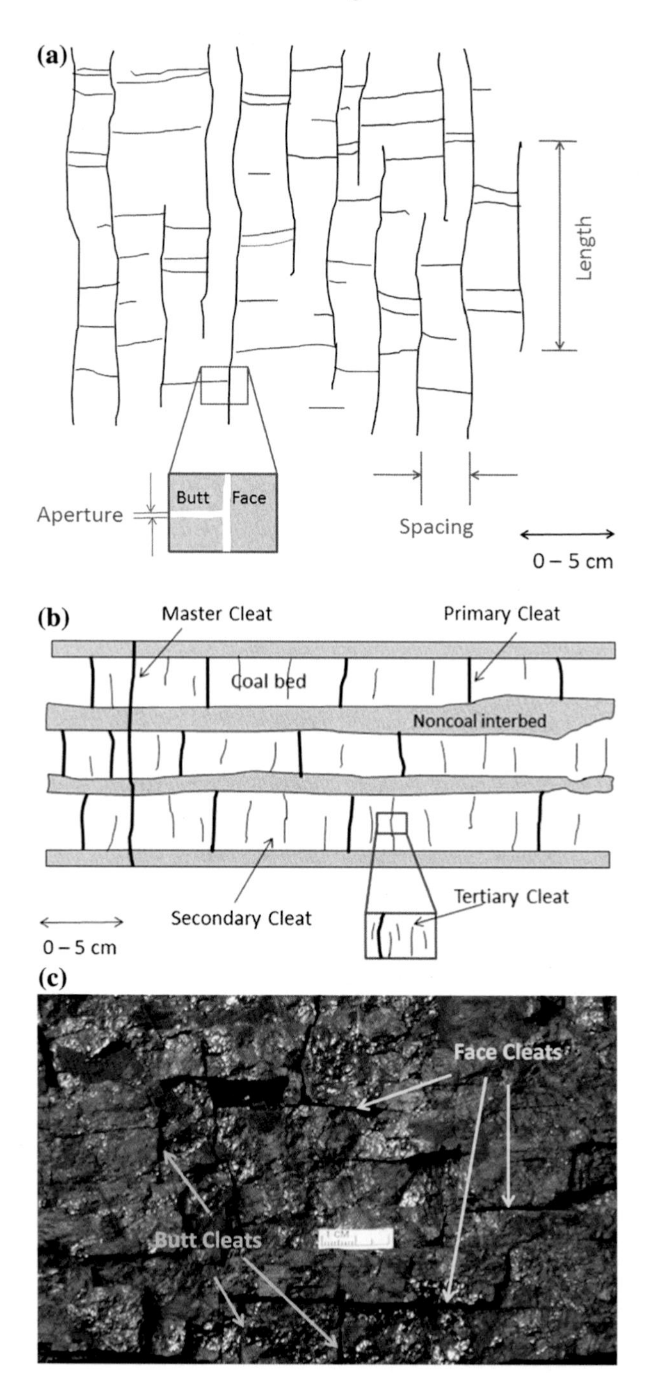

Fig. 2.5 **a** Schematic illustration of coal cleat geometries. Cleat trace patterns are shown in plain view (Redrawn after Laubach et al. 1998). **b** Cleat hierarchies in cross-section view (Redrawn after Laubach et al. 1998). **c** Cleats in an image of coal [From Wolf (2008). Adapted]

2.8 Physical and Chemical Characterization of Coal

In the physical and chemical characterization of coal, two broad categories are distinguished: Proximate Analysis and Ultimate Analysis (Gupta 2007).

Proximate Analyses is the broad analysis of amounts of moisture, volatile matter, ash, and fixed carbon. (Fixed carbon is the material, other than ash, that does not vaporize when heated in the absence of air.) Ash is the inorganic matter present in coal.

Ultimate analysis is the quantitative determination of carbon, hydrogen, nitrogen, sulfur and oxygen within the coal. Sulphur in coal is an important impurity, causing hazardous air pollution as a result of combustion products (sulfur oxides). Three main sources of sulfur are recognized: sulfates, sulfides and organic compounds. In addition, there are chlorine, phosphorus, and the elements that make up the mineral matter (ash).

A number of different parameters have been used to define coal rank, including: moisture and volatile matter content, reflectance, calorific value, organic carbon, hydrogen, and oxygen contents. Unfortunately, no single rank parameter is useful over the entire coalification range (Orem and Finkelman 2005).

The calorific value Q of coal is the heat liberated by its complete combustion with oxygen. This can be measured with a calorimeter.

The inherent fluid in the micro-pores is mostly water. Low-rank coals contain a number of oxygen-functional groups, giving rise to a material with hydrophyllic properties. As a result of this, water content of low-rank coals may be as high as 30–60 wt% on a wet basis (Yu et al. 2013). Anthracite, on the other hand, the highest rank of coal, has a variable water content ranging from 1.5 to 25% (Xu et al. 2013).

Coal is highly complex in nature. The organic matter itself is highly heterogeneous and consists usually of several maceral types with varying physical and chemical structures. The inorganic matter is dispersed randomly in coal, in the form of mineral inclusions and dissolved salts, and is chemically associated with its organic structure (Gupta 2007).

When grinding coal, the material breaks down into a variety of different particles: pure maceral particles, pure mineral particles, and coal particles containing organic matter. The characterization of such a complex material requires the use of more than one analytical technique to accurately predict its behavior during conversion processes such as combustion, gasification, coking, and liquefaction (Gupta 2007).

Also significant for the use of coal is its Ash Fusion Temperature (AFT). The degree of fusion of ash is important in understanding the process of slagging and fouling inside the boilers during combustion of coal (or coal blends). Ash fusion temperatures give an indication of the softening and melting behavior of fuel ash. Ash fusion temperatures are also able to provide an indication of the progressive melting of coal ash to slag (Speight 2013).

References

Bustin RM, Cameron AR, and Grieve DA (1983) Coal petrology: its principles, methods, and applications, vol 3 van Geological Association of Canada short course notes. Victoria, BC, 8–10 May 1983

Chen Y, Mastalerz M, Schimmelmann A (2012) Characterization of chemical functional groups in macerals across different coal ranks via micro-FTIR-spectroscopy. Int J Coal Geol 104:22–33

Coal and Carbon Atlas http://coalandcarbonatlas.siu.edu/coal-macerals/index.php. Visited Aug 2017

Diessel CFK (1992) Coal-bearing depositional systems. Springer Publishers, 721 pp

Encyclopedia Britannica (2018) https://www.britannica.com/science/lignite. Visited June 2018

Gupta R (2007) Advanced coal characterization: a review. Energy Fuels 21(2):451–460

King H (2016) Coal. http://geology.com/rocks/coal.shtml. Visited Sept 2017

Laubach SE, Marrett RA, Olson JE, Scott AR (1998) Characteristics and origins of coal cleat: a review. Int J Coal Geol 35:175–207

Orem WH, Finkelman RB (2005) Coal formation and geochemistry. In: Mackenzie FT (ed) Sediments, diagenesis and sedimentary rocks. Treatise in geochemistry, vol 7. Elsevier, pp 191–217

Petersen H, Rosenberg P, Nytaft HP (2008) Oxygen groups in coals and alginate-rich kerogen revisited. Int J Coal Geol 74:93–113

Spackman W (1958) The maceral concept and the study of modern environments as a means of understanding the nature of coal. Trans NY Acad Sci 20(Series II):411–423

Speight JG (2013) Coal-fired power generation handbook. Scrivener Publishing LLC, 739 pp

Stach E, Mackowsky M-Th, Teichmüller M, Taylor GH, Chandra D, Teichmüller R (1982) Coal petrology, 3 and enlarged edn. Borntraeger, Berlin, Stuttgart, 535 pp

Stopes MC (1935) On the petrology of banded bituminous coal. Fuel 14:4–13

Thomas L (2002) Coal geology. Wiley, Chichester, England, 384 pp

van Krevelen DW (1950) Graphical-statistical method for the study of structure and reaction processes of coal. Fuel 29:269–284

Wikipedia (2017) Aromaticity. https://en.wikipedia.org/wiki/Aromaticity. Visited Aug 2017

Wolf KHAA (2008) Coal—introduction. Course material. Delft University of Technology, 12 pp

World Coal Association (2017) http://www.worldcoal.org/coal/what-coal. Visited Aug 2017

Xu T, Wang D-M, He Q-L (2013) The study of the critical moisture content at which coal has the most high tendency to spontaneous combustion? J Coal Prep Utilization 33(3):117–127

Yu J, Tahmasebi A, Han Y, Yin F, Li X (2013) A review on water in low-rank coals: The existence, interaction with coal structure and effects on coal utilization. Fuel Processing Technology 106, 9–20

Zeng F (2016) Organic and inorganic geochemistry of coal. In: Coal oil shale bitumen heavy oil and peat, vol 1. Encyclopedia of Life Support Systems, Developed under the Auspices of the UNESCO, EOLSS Publishers, Paris, France. Accessed Dec 2016

Chapter 3
South Limburg and Adjacent Areas as a Part of the Former Microcontinent Avalonia

Abstract This chapter gives an overview of the plate tectonic processes which are at the basis of the deposition of coal layers in the South-Limburg area: the plants from which the coal was formed were of tropical nature, and thus the region was once located in a tropical climate. The position of this part of Europe in the tropics is due to the movement caused by plate tectonics. This concerns the evolution of the former microcontinent Avalonia, which is now a part of Western Europe and North America. The formation of the coal layers is also described, and the most important carboniferous plants are mentioned.

Keywords Plate tectonics · Avalonia · Microcontinent · Continental drift · Western Europe · Climate zone in the carboniferous · Formation of coal layers · Carboniferous plants

3.1 Introduction

The lush plant growth that gave rise to the coal layers in South Limburg and the adjacent German and Belgian areas was taking place in a tropical climate. Since the wide acceptance of the concept of continental drift, the tropical climate in Europe in the Carboniferous Period can be addressed from the viewpoint of the location of Western Europe on the face of the Earth in the Upper Carboniferous, a location which was apparently in the tropics. This is explained more in detail below.

Parts of Europe and North America once belonged to the Precambrian and Paleozoic microcontinent Avalonia. In Europe, this nowadays comprises the crust of the northern part of France, the Benelux countries, north-western Germany, north-western Poland, England, Wales, and southern Ireland. Avalonia once was part of the ancient supercontinent Gondwana.

The name Gondwana (also called Gondwanaland) has been given to an ancient supercontinent, which was formed prior to the very large supercontinent Pangea and later became part of it. This has to do with so-called accretionary orogenesis (Cawood and Buchan 2007). The various components of West and East Gondwana were joined, and by the earliest Cambrian (~540–530 Ma), the assembly of Godwana

J. H. L. Voncken, *Geology of Coal Deposits of South Limburg, The Netherlands*, SpringerBriefs in Earth Sciences, https://doi.org/10.1007/978-3-030-18286-1_3

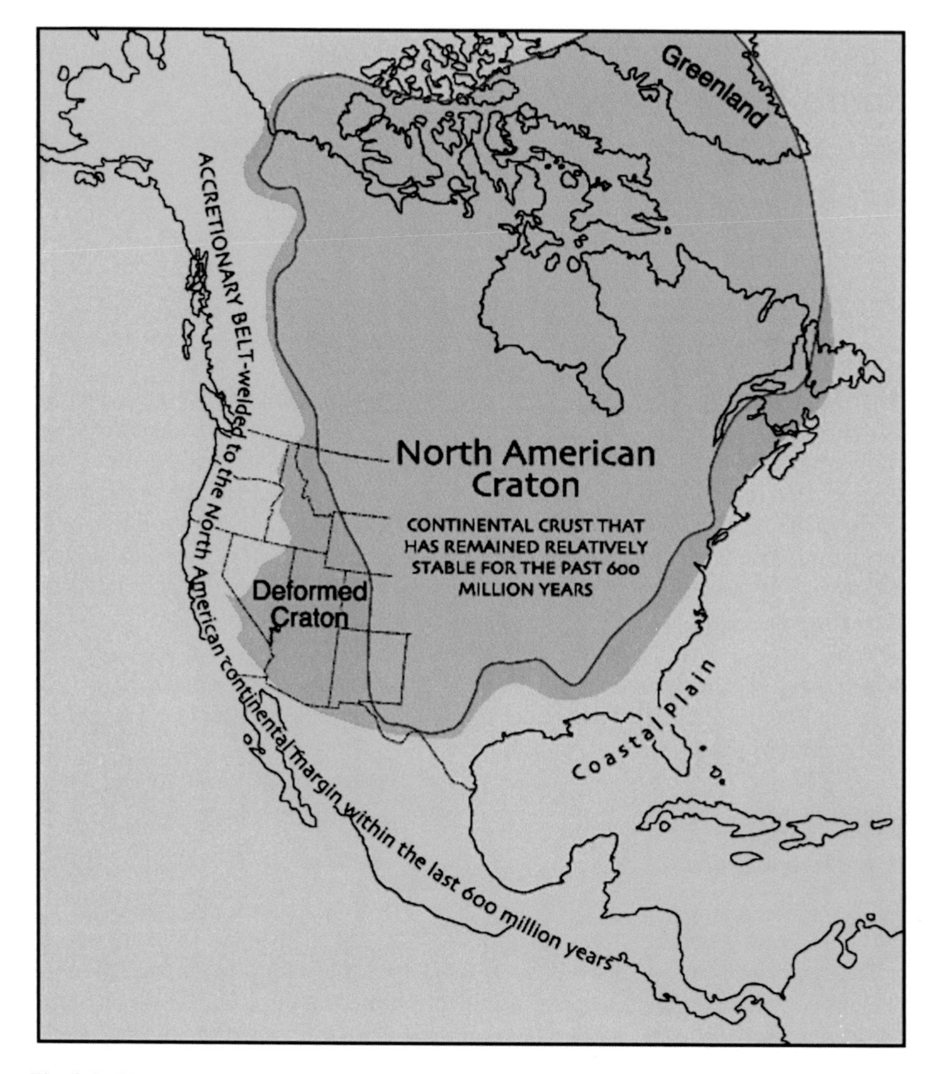

Fig. 3.1 The North American Craton, also referred to as Laurentia (Image courtesy of the U.S. Geological Survey)

was largely complete, except for small oceanic basins in the region of the Kalahari craton (Meert and Liebermann 2008). Cooling and uplift was complete by 490 Ma, in the Late Cambrian (Cawood and Buchan 2007).

Laurentia, consisting mainly of the Precambrian shields of Canada and Greenland, the covered platform and basins of the North American interior, and the reactivated Cordilleran foreland of the Southwestern United States, was formed in the Early Proterozoic period (Hoffman 1988). Laurentia is also referred to as the North American Craton (Fig. 3.1).

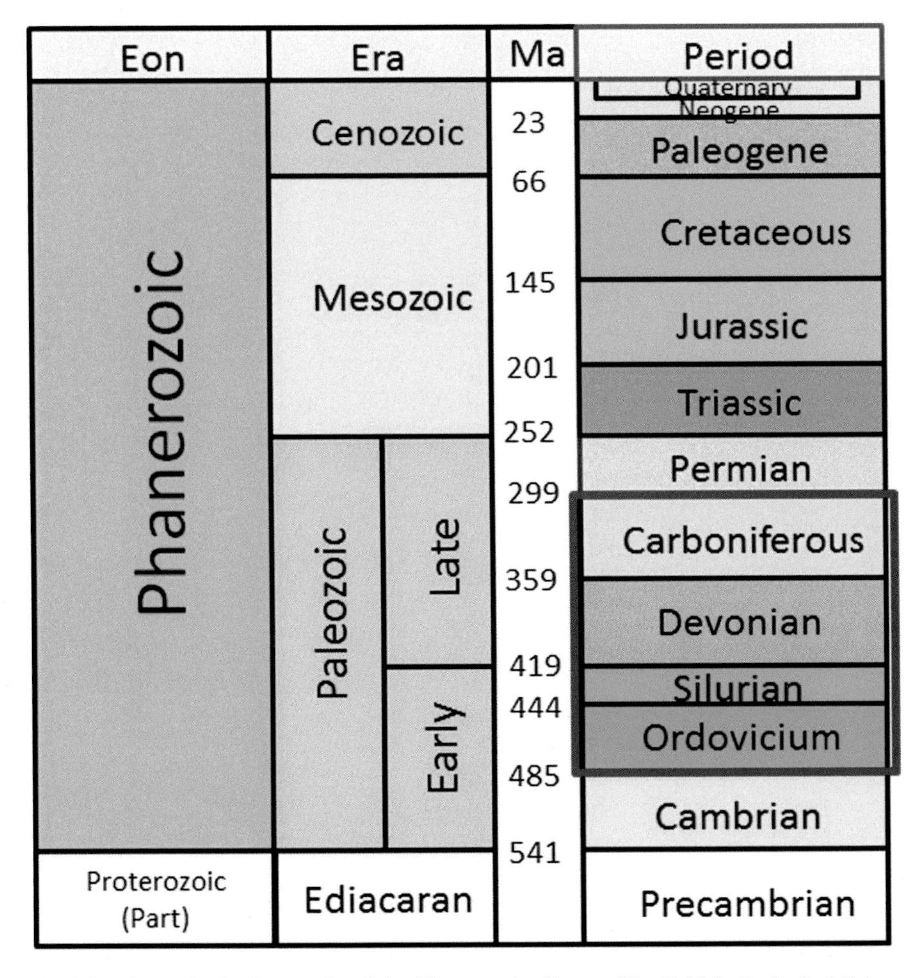

Fig. 3.2 The geologic time scale of the Phanerozoic (*Source* The British Geological Survey. Redrawn). The rectangle indicates the periods mentioned in the text. Ma signifies a time unit of millions of years ago

Below, the geological time scale for the Phanerozoic (i.e., the part of geologic time when there was abundant and differentiated life on Earth) is shown, to provide a reference for the different geological time periods mentioned in the text (Fig. 3.2).

On the northern margin of Gondwana, Avalonia probably developed as a volcanic arc (Fig. 3.3, see also Von Raumer et al. 2003). In the late Darriwilian period (Middle Ordovicium, ±460 Ma ago), Avalonia must already have been detached from Gondwana. By early Silurian times, a widening Rheic Ocean had developed between Avalonia and Gondwana (Cocks et al. 1997).

The paleogeographical evolution of the paleocontinent Avalonia is further shown in Figs. 3.4, 3.5, 3.6 and 3.7. These figures show the changing continental configu-

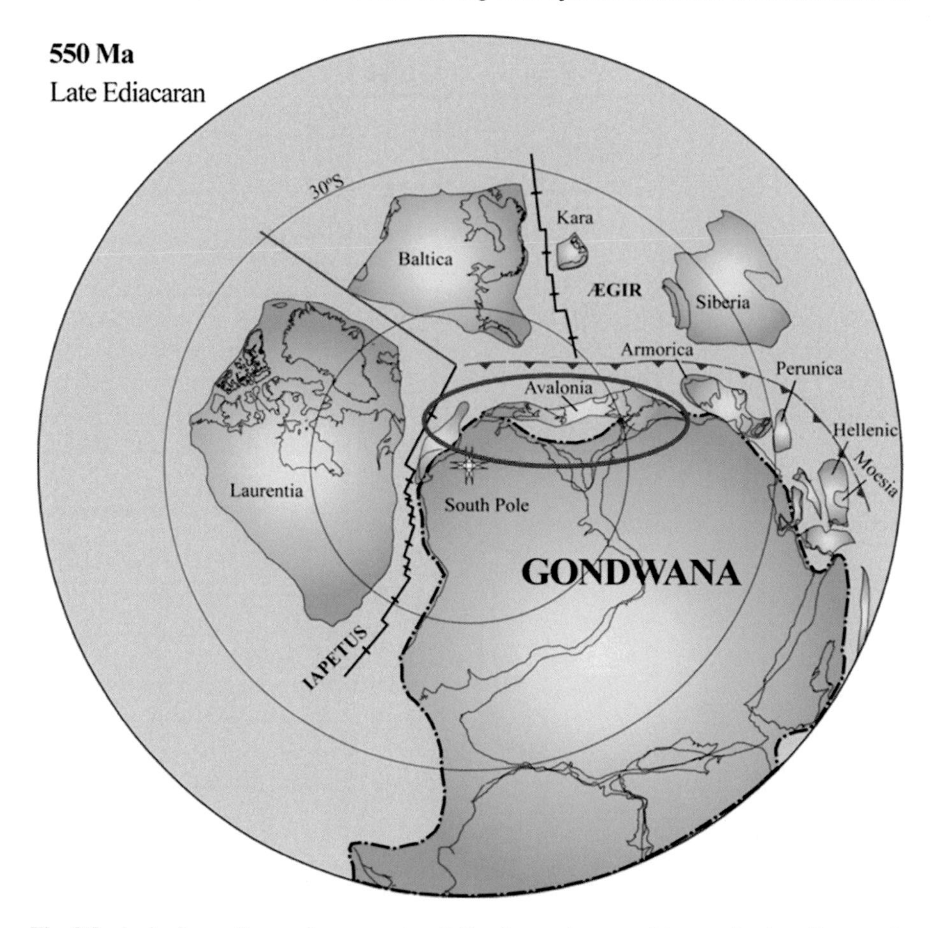

Fig. 3.3 Avalonia on the northern margin of Gondwana (center of image, in the ellipse). The Ediacaran is the last and youngest part of the Precambrium, (543–605 Ma) and is the last period of the Neoproterozoic (Image from Cocks and Torsvik 2006) (Used with permission)

rations from the Ordovician period until the Late Carboniferous. From Fig. 3.7 it is clear that in the Upper Carboniferous, Western Europe (being a part of Avalonia), was in the tropics.

Baltica was a late-early-Palaeozoic continent that now includes the East European craton of northwestern Eurasia (Meisner et al. 1994). Baltica (the Baltic shield) is equivalent with the Fennoscandian shield.

By early Middle Devonian times, faunal evidence indicates that Avalonia was inseparable from Baltica and Laurentia, and merely represented the seaboard margin of the large palaeocontinent Laurussia (Cocks et al. 1997). Compare with Fig. 3.6.

The collision of Avalonia with Baltica thus took place in the Late-Ordovicium. Avalonia and the European Massifs were located in that period at high southerly

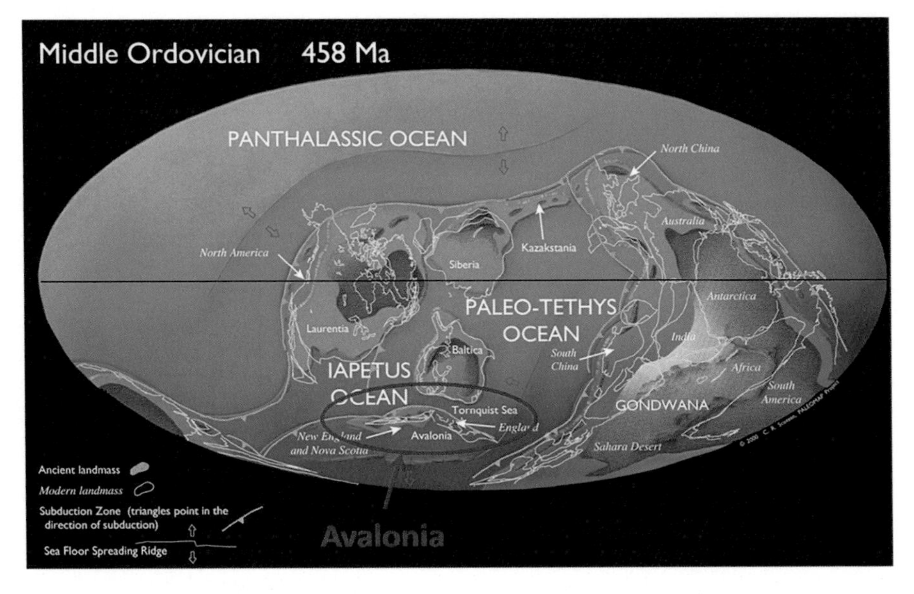

Fig. 3.4 Paleogeography in the Middle Ordovicium (Image from the Geomap Project by R. Scotese. Used with permission)

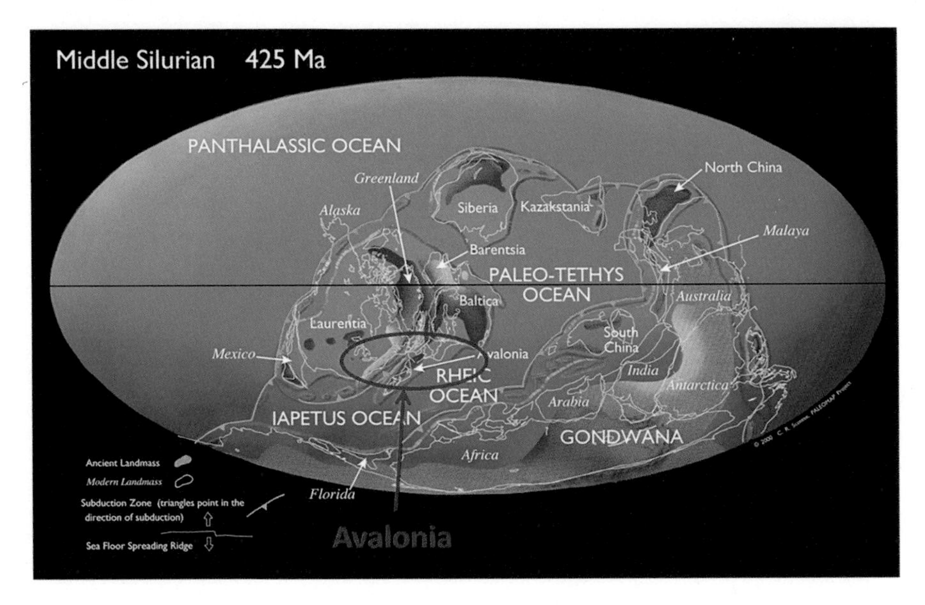

Fig. 3.5 Paleogeography in the Middle Silurian. Avalonia is in the red ellipse (Image from the Geomap Project by R. Scotese. Used with permission)

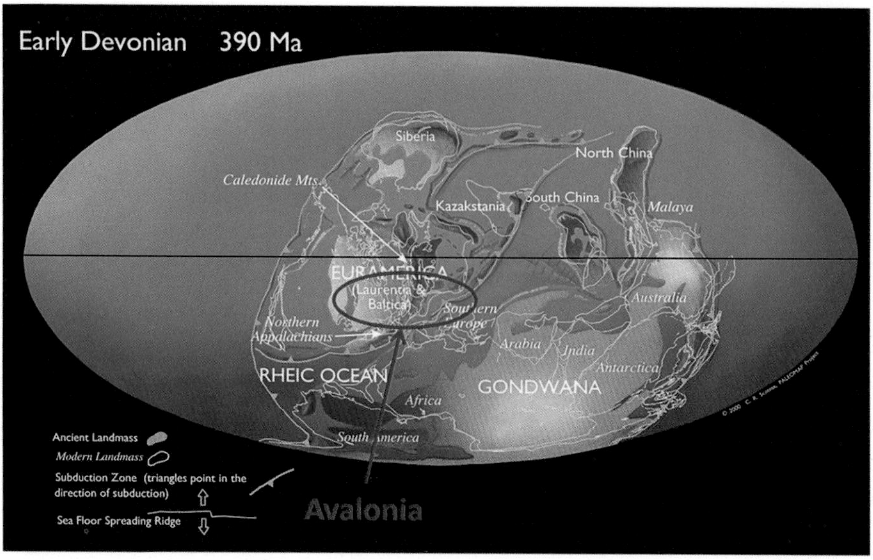

Fig. 3.6 Paleogeography in the Early Devonian. Avalonia is indicated in the ellipse (Image from the Geomap Project by R. Scotese. Used with permission)

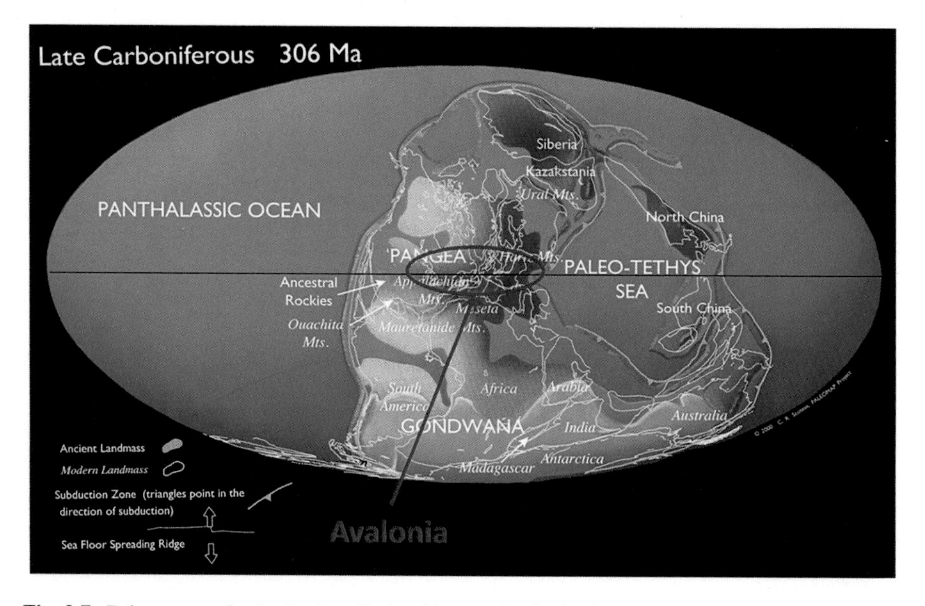

Fig. 3.7 Paleogeography in the late Carboniferous. Avalonia, being part of the not yet complete supercontinent Pangea, is in the ellipse. Avalonia, and therefore Western Europe, was thus on, or near the equator at that time. This explains the tropical climate of those days (Image from the Geomap Project by R. Scotese. Used with permission)

latitudes, just north of the northern margin of Gondwana (Torsvik and Rehnström 2003). See also Fig. 3.3.

Avalonia collided several times: first with Baltica, then with Laurentia, (the Caledonian Orogeny) and finally with Gondwana, (called the Arcadian Orogeny in the US, and the Variscan or Hercynian Orogeny in Europe) (e.g.: Tanner and Meisner 1996; Pharaoh and Carney 2000; Stampli et al. 2002).

The present-day location of the former Avalonia crust in Western Europe is shown in Fig. 3.8. The final completion of the supercontinent Pangea took place in the Early Permian (Ross and Ross 1985). Although it is not completely clear from the paleogeographic reconstruction shown in Figs. 3.6 and 3.7 in the Devonian, and also in the Carboniferous, Avalonia was bordered on one side by open water, although probably not by the Paleotethys Ocean (Domeier and Torsvik 2014). The possibility of islands related to Avalonia is very high, but is usually not depicted in paleogeographical maps (e.g. Cocks and Torsvik 2006).

Rocks that have been deposited on Avalonia can be seen, for instance, in the Belgian Ardennes, and in Wales. An important tectonic unit containing material from Avalonia was the London-Brabant Massif.

3.2 The London-Brabant Massif

The Avalonia microcontinent was not a homogeneously rigid unit. A northern and a southern part were separated by a relatively narrow elongated subsiding zone. It is this zone that evolved into the London-Brabant Massif (Van den Berghe et al. 2014)

The London-Brabant Massif is a WNW–ESE trending structural unit. It consists of a Precambrian crystalline basement, and metasedimentary and volcanic rocks of early Paleozoic age. The London-Brabant Massif is the most easterly part of the Anglo-Brabant Massif (Rijkers et al. 1993).

As a result of the accretion of several microcontinents which formed Laurussia, the London-Brabant Massif became attached to the northern plates, and was repeatedly folded and uplifted during the Early Paleozoic (Middle Cambrian to early Devonian). From Devonian times onward, it acted as a relatively stable platform (Rijkers et al. 1993). The London-Brabant Massif is probably underlain by a granitic body, emplaced at the end of the Ordovician, which is considered the reason that it remained a stable block through time (Kombrink 2008).

The London-Brabant Massif is thought in certain times (especially in the Carboniferous) to have been an island, and is thus termed London Brabant Island. The period from which the island has exercised most economic influence on modern Europe was the Carboniferous. While during the Carboniferous the continent was drifting past the Equator (Fig. 3.7), to the north of the island a basin was formed: the Campine-Brabant Basin. On the margins of this basin paralic conditions[1] prevailed, leading to the successive deposit of thick peat sequences. These peat sequences were,

[1] Paralic conditions: conditions on the landward side of a coast.

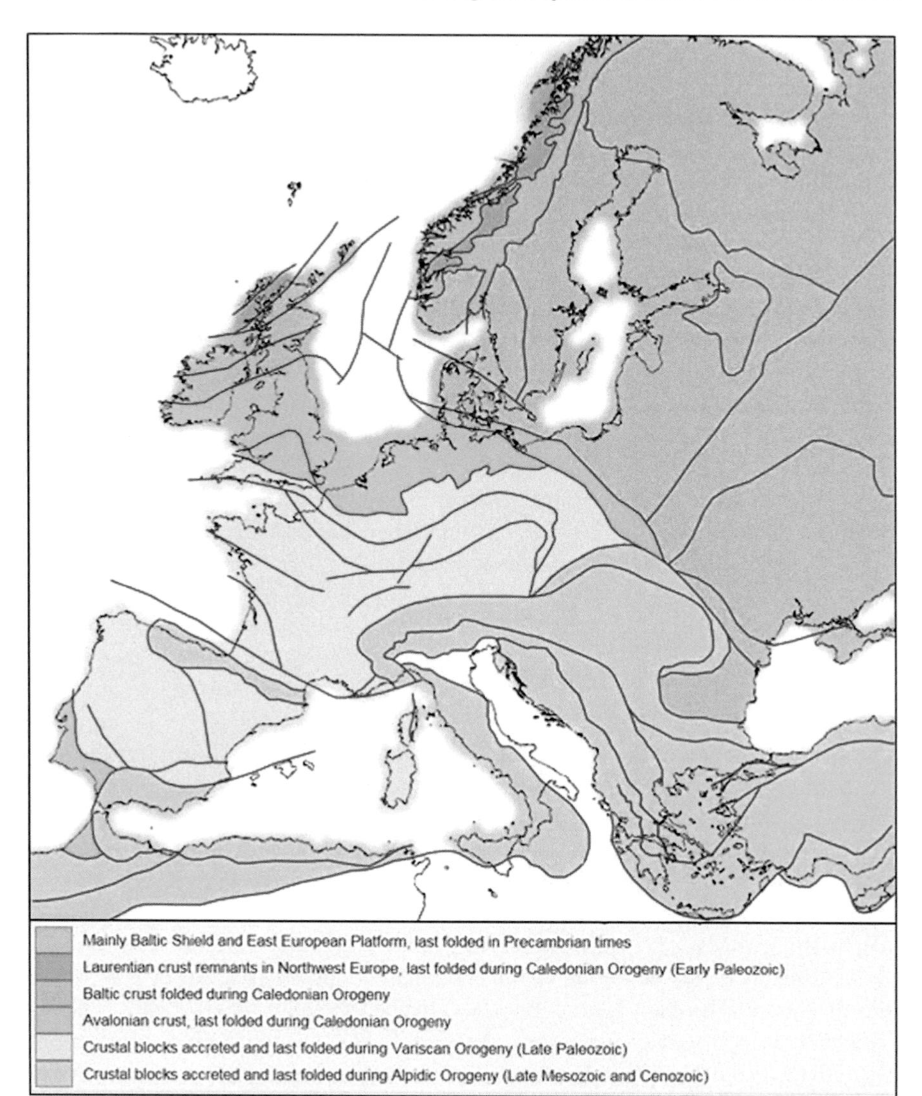

Fig. 3.8 Tectonic sketch map of Europe. Avalonia crust is shown in green (After Wikimedia Commons, adapted. Original image by Woudloper)

by later burial, turned by coalification into thick coal sequences. Out of these marshes the coal fields of Belgium, Germany, and South Limburg were formed (Ziegler 1975; Bless et al. 1980; Rijkers et al. 1993). These coal seams, which are buried deeper in the north of The Netherlands and the under the North Sea, also form the source rocks for the gas found in the southern North Sea, and the adjacent Netherlands and German onshore area (Ziegler 1975) (Fig. 3.9).

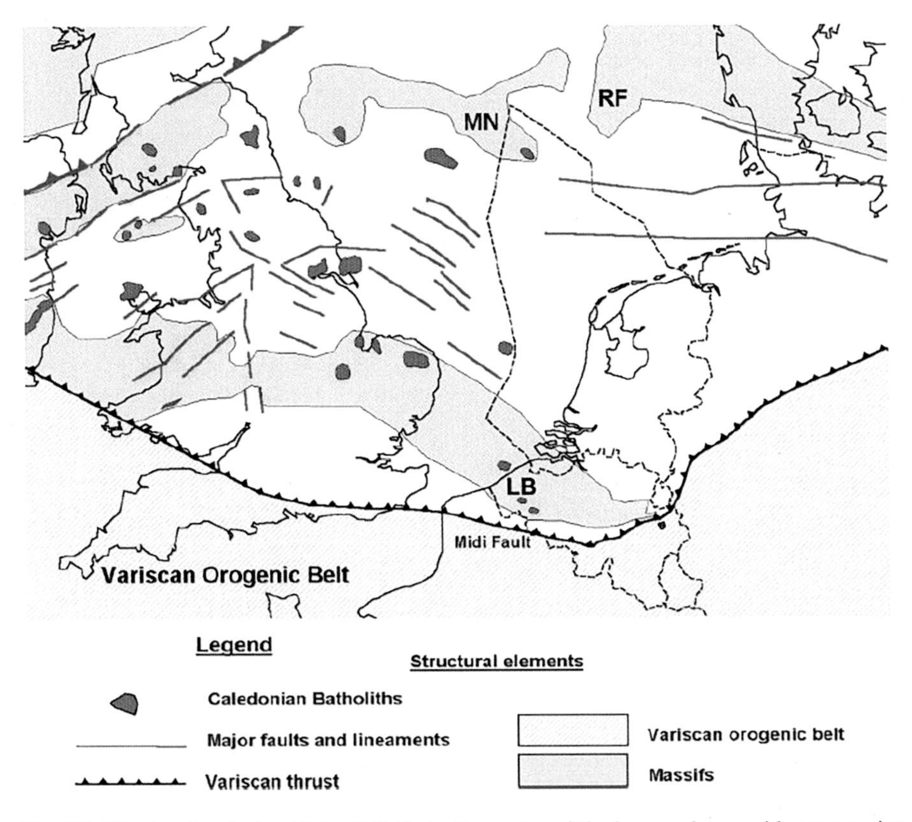

Fig. 3.9 The London-Brabant Massif (LB), in the center of the image, shown with some major tectonic structures in Western Europe. The Variscan Thrust is known in Belgium as the Midi Fault or Faille du Midi (i.e. "Fault of the South") Image from Van Hulten and Poty (2008), used with permission

3.3 Flora and Fauna in South Limburg and Adjacent Areas During the Upper Carboniferous

In the Carboniferous, the flora of South Limburg and adjacent areas was of a tropical nature. This is evident from the position of Western Europe on the globe (see Fig. 3.4): Western Europe is located approximately on the equator.

During the Early Carboniferous era, important groups of plants were the Equisetales (horse tails), the Sphenophyllales (a type of fern), Lycopodiales (a family of plants which also encompasses the modern "witches dance"), the Lepidodendrondales (from the Greek, "scale tree"), which were primitive, vascular, arborescent tree-like) plants, the Filicales (ferns), the Medullosales (a kind of seed ferns) and the Cordaitales (a now extinct type of mangrove trees). During the late Carboniferous period, other groups of plants appeared, including the Cycadophyta (cycads), Callistophytales (also a kind of seed ferns), and the Voltziales (plants belonging to the

Fig. 3.10 Lepidodendron. *Image source* Wikimedia Commons. Original Image by Tim Bertelink

conifers). The most important plants in the coal swamps were Lepidodendron, or Lepidodendrondales, (Encyclopedia Britannica, 2017a), also known as "scale tree", a tree-like plant related to the lycopsids or club mosses, Calamites (belonging to the Calamitaceae, (Encyclopedia Britannica 2017b) and Sigillaria (also belonging to the lycopsids or club mosses) Encyclopedia Britannica 2017c). See Figs. 3.10, 3.11 and 3.12.

Fig. 3.11 Calamites. Reconstruction. Image from Depositphotos.com. Used under license

Fig. 3.12 Sigillaria (*Image source* Wikimedia Commons. Original image by Tim Bertelink)

Lepidodendron could reach lengths up to 40–50 m and trunk diameters of more than two meters. Sigillaria reached similar or slightly smaller sizes (30 m height). Calamites was a so-called horsetail, which, unlike their modern herblike relatives, was a medium-sized tree, which could reach heights of some 30 m (Milsom and Rigby 2004) (Fig. 3.13).

Fauna in the Carboniferous consisted of fauna in the sea and fauna on land. Fauna in the sea consisted, among others, of foraminifera, and attached filter feeders such as bryozoans. The sea floor was dominated by brachiopods. Trilobites became increasingly scarce (UCMP Berkeley 2017). The first terrestrialization of animal life (macroscopic eumetazoans[2]) occurred in the Ludlowian epoch of the Silurian, 425 Ma, while

[2]Eumatozoa are macroscopic animals possessing developed organs.

Fig. 3.13 Reconstruction of a Carboniferous river bank forest (Image by Walter Myers. Used under license)

the first terrestrial vertebrates appeared during the Visean stage (346.7–330.9 Ma) of the Early Carboniferous (Ward et al. 2006).

There is a gap in the evolution of terrestrial vertebrates between 360 and 345 Ma, (Upper part of the Famennien, Devonian, to the upper part of the Tournaisian, Lower Carboniferous) and occurred when environmental conditions were unfavourable for air-breathing, terrestrial animals. This time period is known as Romer's Gap. It is named after Alfred Romer,[3] who observed that few good fossils of tetrapods have been recovered from early Carboniferous deposits (Romer 1956). This gap has been suggested to be due to the low oxygen content of the Earth's atmosphere in those times (Ward et al. 2006).

Oxygen levels in the Earth's atmosphere started to rise in the Upper Devonian (approximately 380 Ma), reaching a peak level of approximately 35% in the latest part of the upper Carboniferous (Berner 1999).

The Carboniferous swamp fauna consisted, among many other varieties, of the first land snails, giant insect-like dragonflies and mayflies. Centipedes and millipedes, spiders and scorpions were also present.

When many animals became adapted to living on land, the amniote[4] egg for reproduction followed. Lizard-like animals such as Hylonomous[5] and Amphibiamus[6], appeared. Anthracosaurs appeared during the Carboniferous, and were quickly

---

[3] Alfred Sherwood Romer, American palaeontologist and biologist (1894–1973).

[4] Amniota: a group of limbed vertebrates that includes all living reptiles (class Reptilia), birds (class Aves), mammals (class Mammalia), and their extinct relatives and ancestors.

[5] An extinct genus of reptile that lived 312 million years ago during the Late Carboniferous period. (Reference: Wikipedia 2018a.)

[6] Amphibiamus is an early temnospondyl. Temnospondyli are a diverse subclass of small to giant tetrapods—often considered primitive amphibians—that flourished worldwide during the Carboniferous, Permian, and Triassic periods. Wikipedia (2018b).

followed by diapsids[7]: The synapsides[8] also made their first appearance (UCMP Berkeley 2017).

Despite this large population of animals,, fossils of animals found in coal deposits are extremely rare. There are two reasons for this:

1. The depositional environment of coal is swamp, and swamps are harsh environments for animal material to survive. Most often, the remains would either dissolve, or be eaten, or be dominated by so many plant remains.
2. To make a swamp into coal requires high temperatures and pressures after burial. This breaks down most organic matter. Plant material survives more often because it is more durable than animal material in this situation, even moreso than bone.

3.4 Thickness of the Coal Layers and Subsidence

The coal deposits of South Limburg and the adjacent Belgian and German areas consist, in general, of rather thin coal seams, alternating with thin shale layers or sandstone layers. Only rarely coal layers of more than a meter thick were formed. Well known thick coal layers from the South-Limburg and Aachen region are the layers with the local names "Grauweck" (average thickness 2.10 m) en "Senteweck" (average thickness 0.75 cm). They were separated by a thin layer of rock (0.25 m). Combined, these layers locally reached a thickness of 3 m (Voncken 2018; Steenkool 1953). Stopes in these layers were sometimes so high that miners could stand up in them. Usually, however, miners had to lie flat and crawl, while mining coal (see Chap. 1, Fig. 1.9).

In the Upper Carboniferous in South Limburg, and in the adjacent Belgian and German regions, paralic conditions prevailed (Ziegler 1975; Delmer et al. 2001), and floods were a regular phenomenon. During these floods, sedimentary layers of sand and silt were deposited, which initially protected the peat that was formed, and eventually became sand and shale layers intercalated with the coal layers. The whole area was subject to subsidence, leading to thick sequences of coal and intercalated sandstone and shale layers. A discussion of the possible subsidence mechanisms is given by Kombrink et al. (2008). On the basis of modelling, these researchers reject the possible explanations of subsidence by flexural subsidence, and lithospheric stretching with associated thermal subsidence, and conclude that the concept of dynamic topography[9] provides the most likely explanation for this phenomenon.

---

[7]Diapsids ("two arches") are a group of amniote tetrapods that developed two holes in each side of their skulls during the late Carboniferous period. The diapsids are extremely diverse, and include all crocodiles, lizards, snakes, tuatara (Sphenodontia), turtles, and dinosaurs (both avian and non-avian). (Reference: Wikipedia 2018c.)

[8]Present-day mammals and their extinct relatives. (Reference: Lauren and Reisz 2018.)

[9]In geodynamics, dynamic topography refers to topography generated by the motion of zones of differing degrees of buoyancy (convection) in the Earth's mantle (Wikipedia 2018e).

They consider that a large amount of subsidence is needed to explain the very thick carboniferous sequences found in the Dutch subsurface (maximum preserved sediment thickness of the Carboniferous succession is up to 6 km). This does not include the area of the South Limburg and Campine coal fields. The paper by Kombrink et al. (2008) focuses on the North Sea area, and the oil and gas-fields which are present there. In the south-Limburg and Campine coal fields, the thickness of the total carboniferous is much less. This is mainly due to erosion during the Early Permian (Brouwer 1972). Lower Permian red beds lie unconformably on the Westphalian strata (van Wijhe et al. 1980; Ziegler 1990; Dinoloket 2018e).

The thickness of the Lower Carboniferous (known as the Lower Carboniferous Limestone Group) is measured from a drill hole, and found to be 953 m, as measured in the drill hole (Dinoloket 2018a). The thickness of the Upper Carboniferous, bearing the once-mined coal layers is measured in a drill hole and found to be in total 1659 m along hole. This is the combined thicknesses of the Baarlo Formation (Dinoloket 2018b), Ruurlo Formation (Dinoloket 2018c) and the Maurits Formation (Dinoloket 2018d). It should be noted that not everywhere in this sampling were all the coal layers present. This is largely due to erosion and movement along fault lines (most important, the Heerlerheide Fault and the Feldbiss Fault).

3.5 Age of the Coal Seams

The coal seams of Western Europe all have an age belonging to the Silesian stage (Wikipedia 2018d), and more specifically to the Westphalian (Delmer et al. 2001). In USA-nomenclature, this is known as the Pennsylvanian Series, and, in more detail, as the Bashkirian, Moscovian and Kasimovian Stages (Fig. 1.8).

The ages in Fig. 1.8 are in Ma and are conform to the International Chronostratigraphic Chart. They have been determined by using a range of geochronologic techniques, including stable and unstable isotopes, palaeomagnetochronology, and astronomical tuning of sedimentary cycles (Cohen et al. 2013).

The depth of burial is estimated from coal maturation data (Dinoloket 2018f).

References

Berner RA (1999) Atmospheric oxygen over Phanerozoic time. Proc Nat Acad Sci USA 96:10955–10957

Bless MJM, Bouckaert IJ, Conil R, Groebsens E, Karsig W, Paproth E, Poty E, Van Steenwinkel M, Streel M, Walter R (1980) Pre-Permian depositional environments around the Brabant Massif in Belgium, The Netherlands, and Germany. Sed Geol 27:1–81

Brouwer GC (1972) The Rotliegend in the Netherlands. In: Schürmann HME, Falke H (eds) Rotligend. International SEdementory petrographical series, pp 34–42

Cawood PA, Buchan C (2007) Linking accretionary orogenesis with supercontinent assembly. Earth-Sci Rev 82:217–256

Cocks LRM, TH Torsvik (2006) European geography in a global context from the Vendian to the end of the Palaeozoic. In: Gee DG, Stephenson RA (eds) European lithosphere dynamics, vol 32. Geological Society, London, Memoirs, pp 83–95

Cocks LRM, McKerrow WS, Van Staal CR (1997) The margins of Avalonia. Geol Mag 134(5):627–636

Cohen KM, Finney SC, Gibbard PL, Fan J-X (2013) The ICS international chronostratigraphic chart episodes 36(3):199–204

Delmer A, Dusar M, Delcambre B (2001) Upper Carboniferous lithostratigraphic units (Belgium). In: Bultynck, Dejonghe (eds) Guide to a revised lithostratigraphic scale of Belgium. Geol Belg 4/1–2:95–103

Dinoloket (2018a) https://www.dinoloket.nl/carboniferous-limestone-group-cl. (DINO = *D*ata en *I*nformatie van de *N*ederlandse *O*ndergrond (*DINO*). *DINO* is the free retrievable database of the Dutch subsurface, made by the Geological Survey of the Netherlands—TNO. The database and website are partly in the Dutch language.) Visited July 2018

Dinoloket (2018b) https://www.dinoloket.nl/baarlo-formation-dccb. Visited July 2018

Dinoloket (2018c) https://www.dinoloket.nl/ruurlo-formation-dccr. Visited July 2018

Dinoloket (2018d) https://www.dinoloket.nl/maurits-formation-dccu. Visited July 2018

Dinoloket (2018e) https://www.dinoloket.nl/node/5173. Visited July 2018

Dinoloket (2018f) https://www.dinoloket.nl/carboniferous. Visited July 2018

Domeier M, Torsvik TH (2014) Plate tectonics in the late Paleozoic. Geosci Front 5:303–350

Ellis D, Stoker MS (2014) The Faroe–Shetland Basin: a regional perspective from the Paleocene to the present day and its relationship to the opening of the North Atlantic Ocean. In: Cannon SJC, Ellis D (eds) Hydrocarbon exploration to exploitation west of Shetlands, vol 397. Geological Society, London, Special Publications, pp. 11–31

Encyclopedia Britannica (2017a) Lepidodendron. https://www.britannica.com/plant/Lepidodendron. Visited Dec 2018

Encyclopedia Britannica (2017b) Calamites. https://www.britannica.com/plant/Calamites. Visited Dec 2018

Encyclopedia Britannica (2017c) Sigillaria. https://www.britannica.com/plant/Sigillaria-fossil-plant. Visited Dec 2018

Encyclopedia Britannica (2018) Carboniferous period. https://www.britannica.com/science/Carboniferous-Period. Visited Apr 2018

Hoffman PF (1988) United Plates of America. The birth of a craton: Early Proterozic assembly and growth of Laurentia. Ann Rev Earth Planet Sci 16:543–603

Houtgast R (2018) De invloed van breuken op het landschap en de loop van rivieren. http://www.geo.vu.nl/~balr/hour/invloed.htm. In Dutch. Accessed Sept 2018

Kombrink H (2008) The Carboniferous of the Netherlands and surrounding areas; a basin analysis. PhD thesis, Utrecht University, The Netherlands, In the Series "Geologica Ultraiectina", No. 294, 184 pp

Kombrink K, Leever KA, Van Wees JD, Van Bergen F, David P, Wong TE (2008) Late Carboniferous foreland basin formation and Early Carboniferous stretching in Northwestern Europe: inferences from quantitative subsidence analyses in the Netherlands. Basin Res 20:377–395

Lauren M, Reisz RR (2018) The tree of life webproject. http://tolweb.org/Synapsida/14845. Accessed Sept 2018

Meert JG, Liebermann BS (2008) The Neoproterozoic assembly of Gondwana and its relationship to the Ediacaran-Cambrian radiation. Gondwana Res 14:5–21

Meissner R, Sadowiak R, Thomas SA (1994) East Avalonia, the third partner in the Caledonian collisions: evidence from deep seismic reflection data. Geol Rundsch 83:186–196

Milsom C, Rigby S (2004) Fossils at a glance. Blackwell Science, Chap. 12, p 107

Pharaoh TC, Carney JN (2000) Introduction to the Precambrian rocks of England and Wales. In: Precambrian Rocks of England and Wales, Geological Conservation Review Series, No. 20, Joint Nature Conservation Committee, Peterborough, 252 pp

Prigmore JK, Butler AJ, Woodcock NH (1997) Rifting during separation of Eastern Avalonia from Gondwana: Evidence from subsidence analysis. Geology 25(3):203–206

Rijkers R, Duin E, Dusar M, Langenaeker V (1993) Crustal structure of the London-Brabant Massif, southern North Sea. Geol Mag 130(5):569–574

Romer AS (1956) The early evolution of land vertebrates. Proc Am Philos Soc 100(3):157–167

Ross CA, Ross JRP (1985) Carboniferous and Early Permian biogeography. Geology 13:27–30

Stampli, GM, von Raumer J, Borel, GD (2002) The Palaeozoic evolution of pre-Variscan terranes: From peri-Gondwana to the Variscan collision. In: Martinez-Catalan JR, Hatcher RD, Arenas R, Diaz Garcia F (eds) Variscan Appalachian Dynamics: The Building of the Upper Paleozoic Basement, Geological Society of America Special Paper, 2002

Steenkool (1953) Bedrijfstijdschrift van de Nederlandse Steenkolenmijnen 11:334–336

Tanner B, Meisner R (1996) Caledonian deformation upon southwest Baltica and its tectonic implications: Alternatives and consequences. Tectonics 15(4):803–812

Torsvik TH, Rehnström EF (2003) The Tornquist Sea and Baltica-Avalonia docking. Tectonophysics 362:67–82

UCMP Berkely (2017) http://www.ucmp.berkeley.edu/carboniferous/carblife.html. University of California, Museum of Paleontology (2017)

Van den Berghe N, De Craen M, Beerten K (2014) Geological framework of the Campine Basin: Geological setting, tectonics, sedimentary sequences. External Report, SCK-CEN, Mol, Belgium, 113 pp

Van Hulten FFN, Poty E (2008) Geological factors controlling early Carboniferous carbonate platform development in the Netherlands. Geol J 43:175–196

van Wijhe DH, Lutz M, Kaasschieter JPH (1980) The Rotliegend in The Netherlands and its gas accumulations. Geol Mijnbouw 59:3–24

Von Raumer J, Stampfli GM, Bussy F (2003) Gondwana-derived microcontinents—the constituents of the Variscan and Alpine collisional orogens. Tectonophysics 365:7–22

Voncken JHL (2018) Private collection

Ward P, Labandeira C, Laurin M, Berner RA (2006) Confirmation of Romer's Gap as a low oxygen interval constraining the timing of initial arthropod and vertebrate terrestrialization. Proc Nat Acad Sci USA 103(45):16818–16822

Wikimedia Commons (2018) https://commons.wikimedia.org/wiki/File:Roerdal_graben_map_NL.svg. Visited Jan 2018

Wikipedia (2018a) Hylonomus. https://en.wikipedia.org/wiki/Hylonomus. Visited Jan 2018

Wikipedia (2018b) Tennospondyli. https://en.wikipedia.org/wiki/Temnospondyli. Visited Jan 2018

Wikipedia (2018c) Diapsid https://en.wikipedia.org/wiki/Diapsid. Visited Jan 2018

Wikipedia (2018d) Silesian (series). https://en.wikipedia.org/wiki/Silesian_(series. Visited Jan 2018

Wikipedia (2018e) Dynamic topography. https://en.wikipedia.org/wiki/Dynamic_topography. Visited Jan 2018

Wikipedia (2018f) https://www.dinoloket.nl/carboniferous. Visited Jan 2018

Ziegler PA (1975) The geological evolution of the North Sea area in the tectonic framework of North-Western Europe. Norges Geol Unders 316:1–27

Ziegler PA (1990) Geological atlas of western and central Europe. Shell Internationale Petroleum Maatschappij. 130 pp. With Enclosures. ISBN 0-444-42084-3

Chapter 4
Later Geological Evolution of South Limburg and Adjacent Areas

Abstract This chapter describes the geological evolution of the South Limburg Coal Mining area from the Carboniferous onwards until the present day. It discusses in detail the sediments deposited during different geological periods, and discusses also the local structural geological situation, which focuses on the Roer Valley Graben, with its still-active faults (Heerlerheide Fault, Feldbiss Fault, and Peelrand Fault Zone). In order to put everything into a larger perspective, the chapter zooms in from an overall picture of the development of the Southern North Sea Basin to the coal-mining area on the south side of the Roer Valley Graben.

Keywords Geological evolution · South-Limburg · Roer Valley Graben · Heerlerheide fault · Feldbiss fault · Peelrand fault zones · Sedimentation Meuse (Maas) river · Quartz sand deposits

4.1 General Picture

To have a better understanding of the geological development of the area around the Roer Valley Graben, *(Dutch: Roerdalslenk)*, it is useful to have some insight into the development of the southern North Sea Basin, and of the tectonic activities in this region of Europe. To facilitate the understanding of readers who are not familiar with the nomenclature and stratigraphy of the different geological periods, a discussion will be given below of the stratigraphy of the Permian and the Mesozoic periods (Triassic, Jurassic, and Cretaceous). Following this, development during the Cenozoic will be described, also including stratigraphic information.

In the Permian, Triassic and Jurassic, a thick sequence of sedimentary rocks has been unconformably deposited on top of the Carboniferous strata, although these strata can only be found in the Roer Valley Graben. The Roer Graben is a part of the north-western branch of the Rhine Graben rift system. It stretches from Euskirchen in Germany to ‘s Hertogenbosch in The Netherlands, traversing a part of Belgium as well (Geluk et al. 1994). Toward the northwest, this Graben system grades into the intraplate basin of the southern part of the North Sea (Zagwijn 1989). See Fig. 4.1.

J. H. L. Voncken, *Geology of Coal Deposits of South Limburg, The Netherlands*, SpringerBriefs in Earth Sciences, https://doi.org/10.1007/978-3-030-18286-1_4

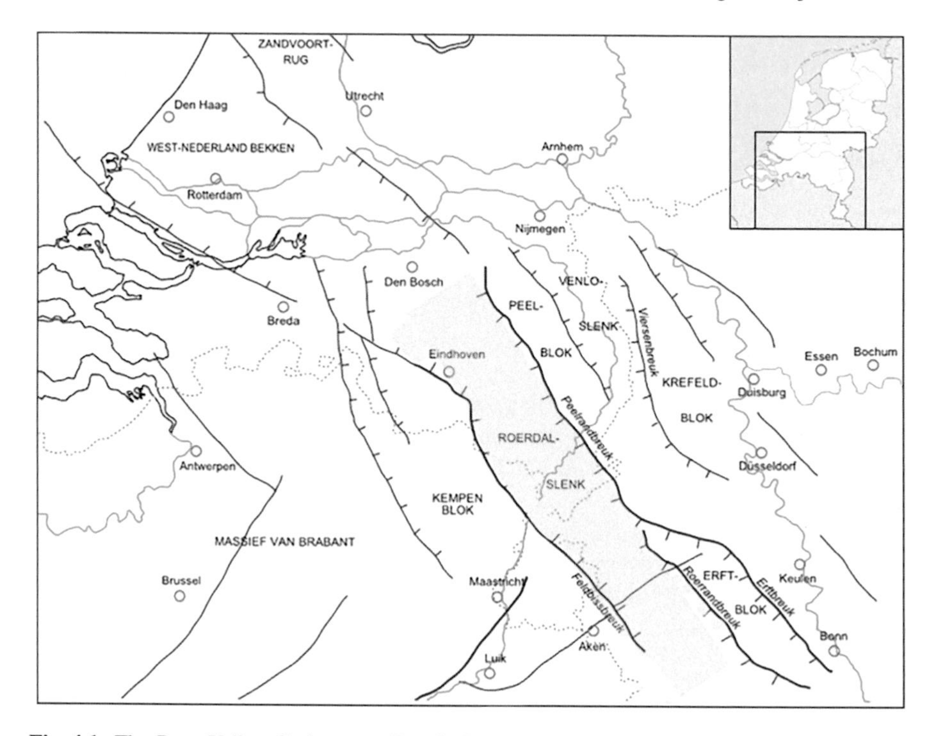

Fig. 4.1 The Roer Valley Graben s.s. (Dutch: Roerdalslenk), (yellow), and its extension to the north-west. Explanation: slenk = graben, blok = horst, bekken = basin (Adapted from Wikimedia Commons (2018a). Original image by "Woudloper". Licensed under CC BY-SA 1.0)

The Roer Valley Graben is essentially of Cenozoic age. Especially during the period of the Alpine Orogenesis, which took place at the plate boundary of Africa and Europe, there were differential movements. Controlling fault patterns, however, already existed during Late Westphalian-Autunian (Upper Carboniferous—Lower Permian) times (Ritsema 1985). These fault patterns show a similar orientation to the fault systems delineating the Cenozoic Roer Valley Graben (Ritsema 1985). The start of the graben formation is considered to have occurred at the time of the beginning of the formation of the Atlantic Ocean. According to Ritsema (1985) and Ellis and Stoker (2014), this happened in the Eocene, although others favor an earlier date (Late Oligocene; Geluk et al. 1994). In NW-directions the Graben system dies out before reaching the present coastline of The Netherlands (Ritsema 1985). Figure 4.1 gives the location of the graben system, and shows it to fade out to the north-west.

Outside the Roer Graben, in South Limburg, sediments from the Upper Cretaceous directly overlie the Carboniferous strata. This is due to the fact that outside the graben and near to its boundary, the sedimentary section is strongly reduced in thickness, or even is completely removed, due to erosion in the Middle Jurassic and through to the end of the Cretaceous. This is due to the Cimmerian tectonic phase. The Cimmerian

("Kimmerian") Orogeny is believed to have begun 200–150 million years ago (much of the Jurassic Period), and to have extended to the Cretaceous and early Cenozoic.

4.1.1 Tectonics

Caused by modifications in the convergence direction of Gondwana and Laurussia, a complex conjugate strike slip fault system developed during the latest Carboniferous—Early Permian (Ziegler 1992). In the Mesozoic, a rift system developed in the North Sea, which was an integral part of the Arctic North Atlantic Mega-rift System (see for instance Ziegler 1989).

The culmination of the evolution of this rift system occurred during the transition from the Paleocene to Eocene, which was the crustal separation between Greenland and northern Europe. In the North Sea, the rifting was never completed, and the North Sea Rift can be seen as a failed arm of the Arctic-North Atlantic rift system (Ziegler 1992). Rifting activity in the North Sea started at the transition from the Permian to the Triassic, and intensified during Middle Jurassic to earliest Cenozoic times. Triassic rifting did not lead to large amounts of crustal stretching, and therefore there was little thermal disturbance of the lithosphere. Activity diminished as the crustal extension between Greenland and Northern Europe concentrated on the zone of the future crustal separation.

4.2 The Permian

The Permian is a geologic period which spans 47 million years, from the end of the Carboniferous Period, 298.9 ± 0.15 million years ago (Ma), to the beginning of the Triassic period 251.902 ± 0.024 Ma. The name is derived from the Russian Oblast (province) Perm. It is the last period of the Paleozoic Era. It follows the Carboniferous Period, and precedes the Mesozoic Triassic Period. The Permian System has been subdivided into three series. These are the Cisuralian, Guadeloupian and Lopingian Series, and together they contain nine stages (Fig. 4.2). The Permian Period was extremely important in terms of its geological history, because it witnessed a process from the greatest icehouse stage to the biggest biological mass extinction stage during the Phanerozoic: the Permian Mass Extinction (also termed "The Great Dying"). This extinction wiped out about 95% of the marine species and 75% of terrestrial species on the Earth at the end of the Permian. There are several theories about its origin, but they will not be discussed here. The Permian world at that time was also dominated by a united supercontinent known as Pangea, surrounded by a global ocean called Panthalassa with a gulf in the eastern part of Pangaea called Palaeotethys (ICS 2018a; see also Chap. 3).

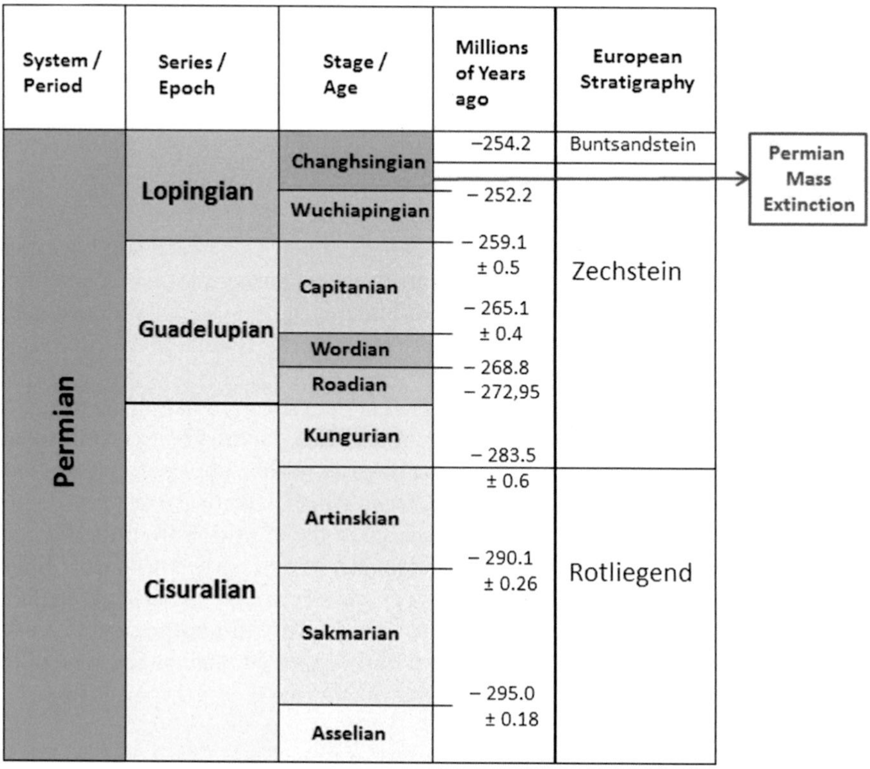

System / Period	Series / Epoch	Stage / Age	Millions of Years ago	European Stratigraphy	
Permian	Lopingian	Changhsingian	–254.2	Buntsandstein	Permian Mass Extinction
		Wuchiapingian	– 252.2	Zechstein	
	Guadelupian	Capitanian	– 259.1 ± 0.5		
		Wordian	– 265.1 ± 0.4		
		Roadian	– 268.8		
	Cisuralian	Kungurian	– 272,95		
		Artinskian	– 283.5 ± 0.6	Rotliegend	
		Sakmarian	– 290.1 ± 0.26		
		Asselian	– 295.0 ± 0.18		

Fig. 4.2 Stratigraphy of the Permian (After the Encyclopedia Britannica (2018a) and the ICS (2018a) (International Commission on Stratigraphy), redrawn). The Asselian is also known as the Autunian. This latter, older name is not much in use anymore

In The Netherlands, the Permian occurs mostly in the subsurface of the northern part of The Netherlands. It consists of two major units: the Rotliegend,[1] a porous sandstone unit, and above that the Zechstein,[2] containing ubiquitous evaporates. These evaporates form an impermeable, plastically deformable cover over the red bed sandstones of the Rotliegend. Degassing of the much deeper-situated carboniferous coals, due to the high temperature at these depths, led to accumulation of methane (CH_4) in the red bed aquifers, sealed off by the Zechstein evaporites. These gas accumulations form the gas deposits of the onshore of The Netherlands and of the North Sea gas fields.

[1]Rotliegend (German) means "red (under)lying."

[2]Zechstein refers to the German term "Zeche Stein," meaning quarry stone.

4.3 The Mesozoic

4.3.1 Triassic

The Germanic Trias was originally defined from outcrops in central Germany, where it was subdivided into Buntsandstein (Lower), Muschelkalk (Middle) and Keuper (Upper). This tripartite subdivision was made using lithostratigraphic criteria, although it has often also been used as a chronostratigraphic subdivision. The term Trias is derived from the three rock sequences found in central Germany, as mentioned above, and means literally "threefold division."

The cyclic deposition of marine evaporates, which was common during the Permian, ended at the onset of the Triassic, because the sea withdrew from the Northern and Southern Permian basins. The conditions thus became terrestrial. Initially, a wide-spread uniform subsidence occurred, but at the end of the Lower Triassic a large domal swell formed, which developed for the most part in The Netherlands onshore (Dinoloket 2018a).[3] The Netherlands Swell was bordered in the south of The Netherlands by the basin of the Roer Valley Graben.

Initially, there was fine-grained sedimentation under the existing lacustrine conditions (Lower Buntsandstein Formation). A cyclic alteration of sandstones and claystones with a strong fluvial influence then followed, forming the main part of the Buntsandstein. At the end of the lower Triassic, there was considerable uplift, leading to erosion of the previously deposited formations (Dinoloket 2018a). On top of this, fine-grained sandstones were deposited, followed by claystones and evaporites of the Röt Formation. The formation forms the upper part of the Bundsandstein. During the Anisian and Early Ladinian (Middle Trias), carbonates of the Muschelkalk Formation were deposited under marine conditions.

In late Ladinian to Norian times (Fig. 4.3), clastic sediments and evaporites were deposited, belonging to the Keuper Formation. These deposits occurred in a continental setting. In the Rhaetian, marine conditions returned (Dinoloket 2018a).

4.3.2 Jurassic

During the Permian and the Triassic, the climate had been an arid continental climate. Widespread red bed and evaporate deposits occurred. At the end of the Triassic, the depositional regime changed considerably. From the Latest Triasic (Rhaetian), the environment became an open marine system (Dinoloket 2018b). In this basin, which encompassed most of present day Western and Central Europe (Ziegler 1992), a thick series of marine fines and marls was deposited. Locally a thinner succession, sometimes with iron-containing layers, was deposited. From the Hettangian until

---

[3]Dinoloket stands for DINO—Data en Informatie van de Nederlandse Ondergrond. (English: Data and Information of the Dutch Subsurface), and Loket (English: public counter, or entry site).

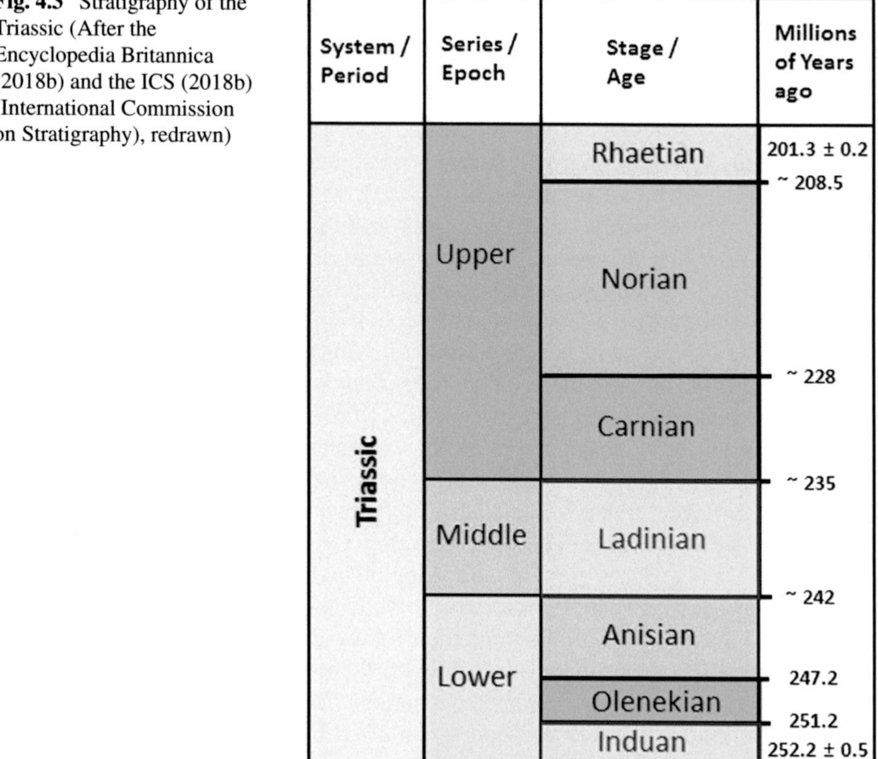

Fig. 4.3 Stratigraphy of the Triassic (After the Encyclopedia Britannica (2018b) and the ICS (2018b) (International Commission on Stratigraphy), redrawn)

the Toarcian (uppermost Lower Jurassic, Fig. 4.4), quiet, open-marine depositional conditions were common in the whole of Western Europe. During the early Toarcian (Fig. 4.4) locally-closed basins with anoxic conditions occurred. From the Late Aalenian to Bathonian uplifting occurred, as a result of the so-called "Mid-Cimmerian" tectonic phase (Ziegler 1992). In several basins, including the basin that was later to become the Roer Valley Graben, deposition of marine sandy carbonates and marls continued into the Callovian and, locally, into the Oxfordian (Dinoloket 2018b). See Fig. 4.4.

4.3.3 *Cretaceous*

The stratigraphy of the Cretaceous is shown in Fig. 4.5. During the lower Cretaceous, a transgression occurred. In the Roer Valley Graben, highly differential subsidence patterns continued at least throughout the Valanginian. In the Roer Valley Graben, fluvial-plain equivalents are found. During the rest of the Lower Cretaceous, the transgression in general proceeded. During the latest Albian (Fig. 4.5) the influx of

Fig. 4.4 Stratigraphy of the Jurassic (After the Encyclopedia Britannica (2018c) and the ICS (2018c) (International Commission on Stratigraphy), redrawn)

System / Period	Series / Epoch	Stage / Age	Millions of Years ago
Jurassic	Upper	Tithonian	145.0 ± 0.8
		Kimmeridgian	152.1 ± 0.9
		Oxfordian	157.3 ± 1.0
	Middle	Callovian	163.5 ± 1.0
		Bathonian	166.1 ± 1.2
		Bajocian	168.3 ± 1.3
		Aalenian	170.3 ± 1.4
	Lower	Toarcian	174.1 ± 1.0
		Pliensbachian	182.7 ± 0.7
		Sinemurian	190.8 ± 1.0
		Hettangian	199.3 ± 0.3
			201.3 ± 0.2

fine-grained, clastic material into the marine environment, which covered most of The Netherlands, suddenly diminished. This was due to a short-term regression, which was followed by a large-scale transgression. Widespread carbonate deposition started at the beginning of the Cenomanian (Fig. 4.5), and was to last until the end of the Maastrichtian, and even into the earliest Paleocene (Fig. 4.6). In the South Limburg area, bioclastic limestones can befound, dating from the Maastrichtian (Dinoloket 2018c).

The Triassic and Jurassic sediments have been eroded in the South Limburg area during Cretaceous times. This is due to the Cimmerian tectonic phases.[4] In the Roer Valley Graben, however, sediments from these ages have been preserved. During the late Cretaceous and earliest Tertiary, subsidence of the area occurred again, accompanied by a general transgression. This led, among other effects, to deposition of rocks belonging to the Chalk Group (Dinoloket 2018d). The Chalk Group is not present in the Roer Valley Graben (Duin et al. 2006, Fig. 3b).

The Chalk Group is a lithostratigraphic unit, known from the late Cretaceous limestone succession in southern and eastern England. (e.g. The Strait of Dover,

[4]The Cimmerian ("Kimmerian") Orogeny is believed to have begun 200–150 million years ago (lasting during much of the Jurassic Period), and to have extended to the Cretaceous and early Cenozoic. It is known for creating ancient fold belts, mainly in South Asia.

System / Period
Series / Epoch
Stage / Age
Millions of Years ago
Cretaceous – Paleogene Mass Extinction
Cretaceous
Late
Early
Maastrichtian
Campanian
Santonian
Coniacian
Turonian
Cenomanian
Albian
Aptian
Barremian
Hauterivian
Valanginian
Berriasian
66.0
72.1
83.6
86.3
89.8
93.9
100
113
126
131
134
139
145

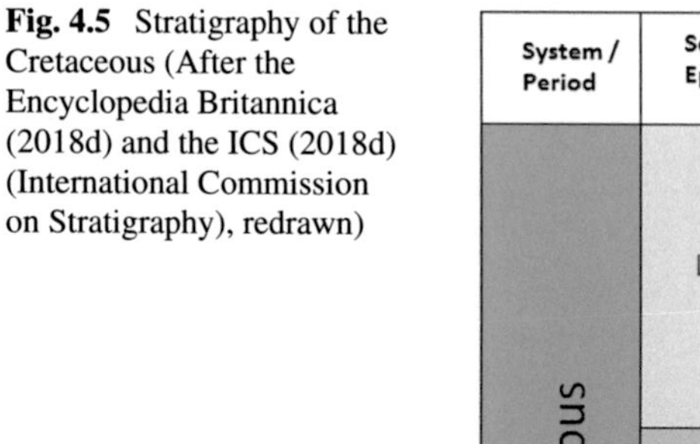

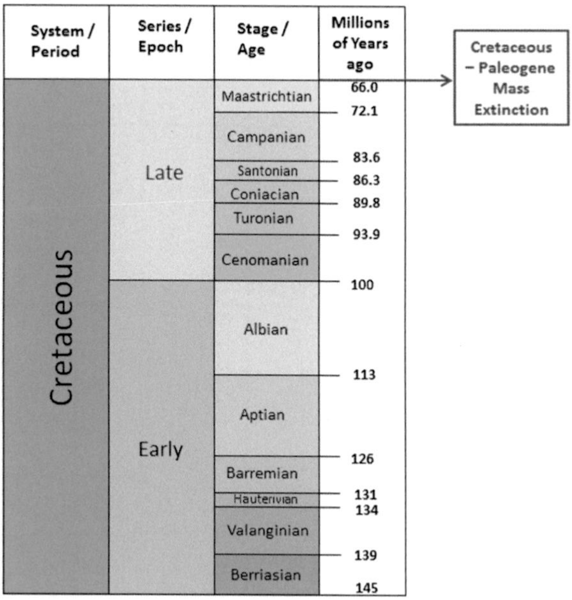

Fig. 4.5 Stratigraphy of the Cretaceous (After the Encyclopedia Britannica (2018d) and the ICS (2018d) (International Commission on Stratigraphy), redrawn)

System / Period
Series / Epoch
Stage / Age
Millions of Years ago
Paleogene
Oligocene
Eocene
Paleocene
Chattian
Rupellian
Priaborian
Bartonian
Lutetian
Ypresian
Thanetian
Selandian
Danian
23.03
28.1
33.9
37.8
41.2
47.8
56.0
59.2
61.6
66.0

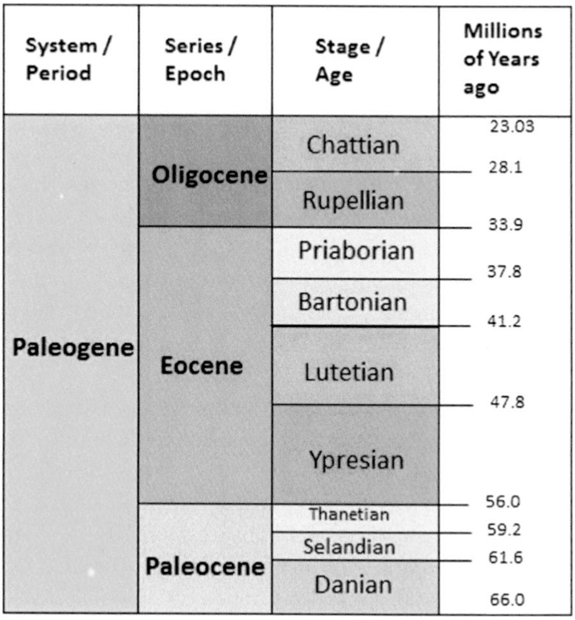

Fig. 4.6 Stratigraphy of the Paleogene (After the Encyclopedia Britannica (2018e) and the ICS (2018e) (International Commission on Stratigraphy), redrawn)

which separates England from France and the European continent, and which connects the English Channel and Atlantic Ocean with the North Sea). In the uttermost south of South Limburg, near the village of Epen, the Upper Cretaceous sediments unconformably overlie outcropping Carboniferous rocks from the Namurian stage (326–313 million years ago).

However, similar rocks occur within a wide area of northwest Europe. The Dutch and Belgian equivalents of the Chalk Group (Dutch: Krijtkalk-Groep; Belgian: Krijt-Groep) are basically continuous, and crop out as a subhorizontal, slightly northwest-dipping layered sequence, in a belt extending from the German city of Aachen to the Belgian city of Mons, where they join Cretaceous deposits of the Basin of Paris. The chalk of the upper Cretaceous of South Limburg is described in detail by Felder and Bosch (2000).

Outside of the graben and near to its boundary, the sedimentary section is strongly reduced in thickness, or is even completely removed, due to erosion (first during the Middle Jurassic and later at the end of the Cretaceous). As mentioned before, this is due to the effects of the Cimmerian tectonic phase. Outside of the graben, Upper Cretaceous sediments directly overlie the Carboniferous strata (Geluk et al., Fig. 4).

4.4 The Roer Valley Graben

According to Ritsema (1985) and Ellis and Stoker (2014), the start of the graben formation occurred in the Eocene, although others favor placing it at an earlier date (Late Oligocene; Geluk et al. 1994).

The Roer Valley Graben is named after the Roer River, a right-bank tributary of the Maas (Meuse).

4.5 Cenozoic Evolution of the Roer Valley Graben

During the Cenozoic, the Roer Valley Graben was affected by two periods of inversion, named the Laramide phase (earliest Paleogene) and the Pyrenean phase (late Eocene—Early Oligocene), and also by continuous subsidence, since the beginning of the Oligocene (Michon et al. 2003).

Most of the Triassic structures were reactivated during the Late Cretaceous and Miocene evolutions (Michon et al. 2003). Although different stress fields were causing faulting, existing NW–SE fault orientations (dating from Variscan times) were reactivated.

Subsidence analysis has revealed two Cenozoic periods of subsidence: Late Paleocene and Oligocene—Quaternary, separated by a hiatus during Eocene time, resulting from a period of erosion during the Late Eocene (Michon et al. 2003).

During the first phase of subsidence (Late Paleocene), the tectonic subsidence rates indicate a different evolution having taken place between the south-eastern part of the Roer Valley Graben, and the Peel Block, and also the north-western part of the Roer Valley Graben. Tectonic activity was, however, mainly restricted to the graben (Michon et al. 2003).

The second phase of subsidence started in the early Oligocene, and was still active during the Quaternary. Initially, the Roer Valley Graben and the Peel Block subsided with similar tectonic subsidence rates, but this parallel evolution stopped at the beginning of the Late Oligocene. At that time, the tectonic subsidence decreased in the north-western part of the graben and in the Peel Block, but strongly accelerated in the south-eastern part of the graben. At the transition of the Oligocene to Miocene, there was a regressive phase, after which the evolution of the north-western and south-eastern parts became similar to each other. Although in earlier times the Peel Block also subsided, from the period of the Oligocene-Miocene transition, movement became restricted to the graben. Tectonic subsidence rates increased during the Miocene-Pliocene transition (Michon et al. 2003).

During the Quaternary, subsidence rates changed rapidly, although Michon et al. (2003) point out that this could have been caused by the averaging of several small periods with strong subsidence with periods of quiescence over the course of several millions of years (Paleogene times), as compared to the more recent periods, where one period with a lot of activity is encountered. On the other hand, an alternative explanation could be found in the comparison of the amount of tectonic and total subsidence, thus taking into account compaction and isostatic processes.

4.6 Miocene Deposits in the Roer Valley Graben

In Miocene times, in the South Limburg area, layers of pure to very pure quartz sand were deposited. These are considered to be dune and shoreline sands. Most of The Netherlands was covered by the sea. South Limburg, and especially the coal mining area, formed the shore of this sea, which comprised dunes and sandbars, but also included swamps, where peat was deposited. This peat formed the source material for later lignite layers (Geologie van Nederland 2017a).

The brown coal and quartz sand in South Limburg is present near the surface, in a thin strip of land between the Heerlerheide Fault and the Feldbiss Fault (Fig. 4.7).

These quartz sand and brown coal layers belong to the Breda Formation, which consists mainly of fluviatile quartz sand and gravel, alternating with lignite layers. The sand quarries shown in Fig. 4.7 belong to the Heksenberg Member of the Breda Formation. The age of the Breda formation is late Oligocene to early Pliocene (Westerhof 2003). See also Fig. 4.8. This means that the quartz sand layers belong, for the most part, to the Miocene age. In the Heerlen-Brunssum-Landgraaf area of South Limburg, deposits of the Kiezeloölite Formation (a Dutch name, translated here into English as "Pebble-Oölite Formation") are found on top of the Breda Formation. The deposits are of Pliocene age (Dinoloket 2018e), and include—among others—the Waubach Member, consisting mainly of gravel and coarse sand (420–2000 μm), and also the Brunssum Member, which consists of clay, which may range from being

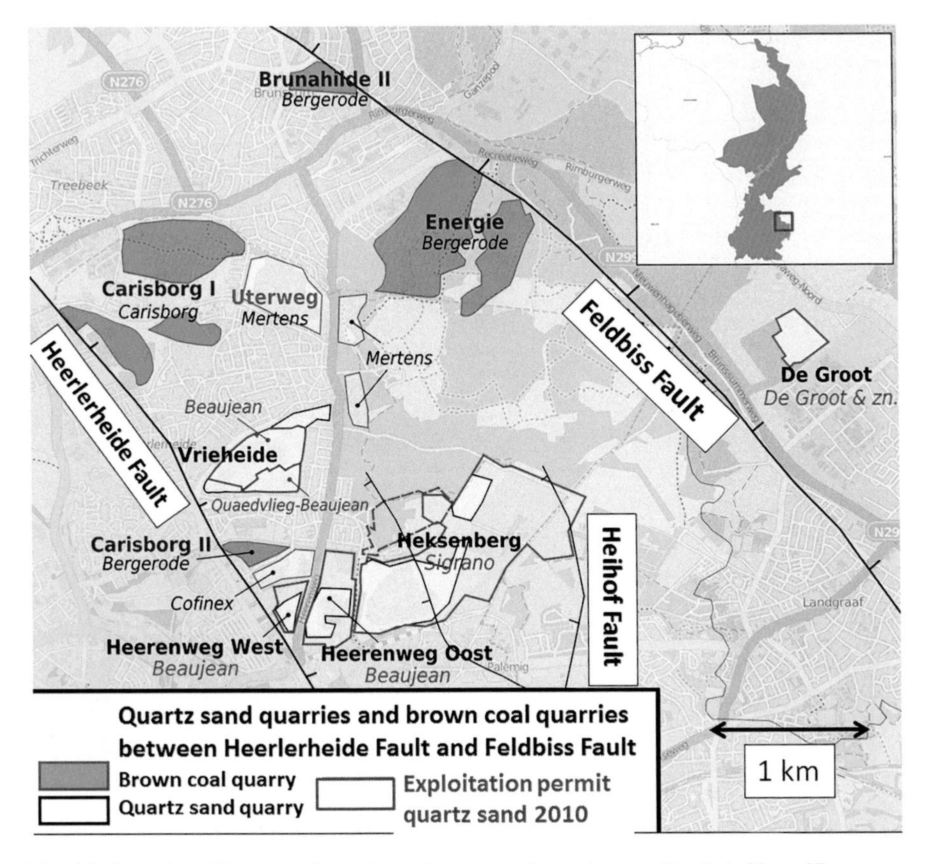

Fig. 4.7 Location of brown coal quarries and quartz sand quarries near Heerlerheide and Brunssum. The Beaujean and Quadflieg-Beaujean quarries do not exist anymore. During the 1990s, they were filled with debris from the waste dump of the coal mine Oranje Nassau 1, and there now remains just a strip of flat land. (*Image source* Wikipedia (2018a). Original image by Hans Erren. Adapted. License CC BY-SA 3.0)

weakly sandy to strong silty, and which also contains intercalations of coarse sand and brown coal (lignite) (Doppert et al. 1975). In the basal part of the formation, quartz sand may occur.

The quartz sand layers vary from 5 to 50 m thickness. Between the towns of Heerlen and Brunssum, there were, in the past, up to nine quarries of different size. They were exploited by a small number of companies (Beaujean, De Groot, Mertens, Conifex). Presently there is only one remaining, owned by the Belgian company Sibelco.[5] The predecessor was called "Sigrano."[6]

[5]In full: Sablières et Carrières Réunies (SCR) Sibelco.

[6]The name "***Sigrano***", is (probably) an acronym of "***Si***licium ***Gran***ules ***O***ranje Nassau," as the quartz sand exploitation was originally started by coal mining company Oranje Nassau Mines. A "granule," in geology, signifies a specified particle size of 2–4 mm.

Fig. 4.8 Stratigraphy of the Neogene (After the Encyclopedia Britannica (2018f) and the ICS (2018f) (International Commission on Stratigraphy), redrawn)

System / Period	Series / Epoch	Stage / Age	Millions of Years ago
Neogene	Pliocene	Piacenzian	2.58 – 3.600
		Zanclean	3.600 – 5.333
	Miocene	Messinian	5.333 – 7.246
		Tortonian	7.246 – 11.63
		Serravalian	11.63 – 13.82
		Langhian	13.82 – 15.97
		Burgdalian	15.97 – 20.44
		Aquitanian	20.44 – 23.03

The lignite (brown coal) was mined during the 20th Century in quarries in Graetheide,[7] Brunssum, Heerlerheide, Eygelshoven and Haanrade (De Mijnen 2018; Heerlen Vertelt 2018).

4.7 Quaternary Evolution of the South Limburg Area

The Quaternary is subdivided into two epochs, the Pleistocene and the Holocene. During the Pleistocene, Northern Europe has been affected by several glaciations.

4.7.1 Pleistocene

Contrary to the north of The Netherlands, where there has been land-ice during the Elster and Saale glaciation (respectively 420,000–470,000 and 130,000–380,000 years ago), the South Limburg area was never covered by an ice sheet. In the South-Limburg area, during the Pleistocene, loess was deposited. Loess is a clastic, predominantly silt-sized sediment that is formed by the accumulation of wind-blown dust (Doeglas 1949; Wikipedia 2018b) (Fig. 4.9).

[7]Graetheide is a township near Sittard, The Netherlands. Eygelshoven and Haanrade are former villages, now belonging to the municipality of Kerkrade, The Netherlands.

Fig. 4.9 Stratigraphy of the Pleistocene (After the Encyclopedia Britannica (2018g) and the ICS (2018g) (International Commission on Stratigraphy), redrawn)

System / Period	Series / Epoch	Sub Series	Stage / Age	Millions of Years ago
Quaternary	**Pleistocene**	Upper	Tarantian	0.0117 – 0.126
		Middle	Ionian (Chibanian)	0.126 – 0.781
		Lower	Calabrian	0.781 – 1.80
			Gelasian	1.80 – 2.58

This happened during the Weichselian and Saalian (de Gans 2007). As loess is an eolian product, its mineralogical and mechanical composition is markedly different from the underlying Cretaceous and Tertiary deposits, and also from the Pleistocene terrace deposits, but resembles that of the deposits in the northern part of The Netherlands (Doeglas 1949). The loess (occurring in layers of 2 m, varying to 20 m) overlies coarse-grained Quaternary fluvial sediments, Paleogene sands, and Late Cretaceous limestone (Imeson 2012).

The fluviatile sediments are deposited by the river Meuse (Maas), which, contrary to today, followed a more easterly course (Fig. 4.11). The oldest traces of the river Meuse are found in the sediments of the Upper Miocene. These deposits are known as the "Kiezeloölietformatie" (Pebble-Oölite Formation), are also known as the Formation of Waubach, which was deposited as an alluvial fan (Bosch 1981). The present Ardennes did not exist at that time, but formed—together with South Limburg—a vast peneplain. Parts of this peneplain are still recognizable in the high areas near the villages of Noorbeek, Epen and Slenaken, the plateaus of Margraten and Kunrade, the plateau of the town of Vijlen, the Vaalserberg ("Mountain of Vaals", elevation 322.4 m above N.A.P.[8]) and the bordering Belgian areas (Bosch 1981) (Fig. 4.10).

At the end of the Tertiary, at the boundary Pliocene-Pleistocene, the uplift of the Ardennes and the Rhenish Massif[9] (German: "Rheinishes Schiefergebirge"; Dutch: "Rijn-Leisteen Plateau") took place (Bosch 1981). This uplift had a large influence on the South Limburg area. The area south of the Feldbiss fault was uplifted, whereas the area located north of this fault (the Roer-Graben) sank. In the Roer Graben, the remnants of the "Pebble-Oölite Formation" were preserved. Elsewhere, they were eroded away.

The oldest morphologically recognizable valley of the Meuse (Maas) River stretches from Eijsden, Noorbeek, Gulpen, Simpelveld, Kerkrade to Jülich in Germany, where the Meuse ended in the Rhine. This is the so-called valley of the East-Meuse (Bosch 1981). See Fig. 4.11. In the North, this valley was bordered by a high

[8]N.A.P. = Normaal Amsterdams Peil (English: "Normal Amsterdam Level"). This is the Dutch reference level. N.A.P. = 0 m means that the elevation is equal to average level of the North Sea.

[9]Sometimes also translated as Rhenish Slate Mountains.

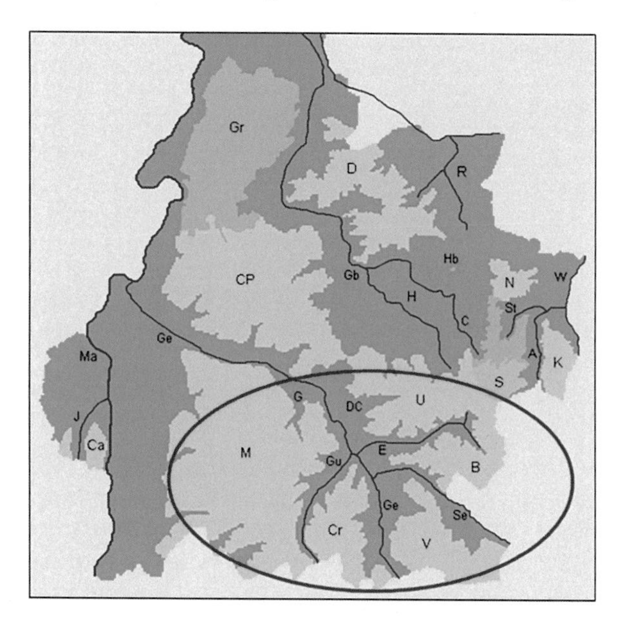

Fig. 4.10 Map with plateaus and valleys of South Limburg, The Netherlands. The oval gives approximate position of the areas mentioned in the text above. After Wikipedia, License CC0 1.0 Universal (CC0 1.0) Public Domain Dedication, adapted. *Legends* **B** Plateau of Bocholtz; **Ca** Plateau of Caestert; **CP** Central Plateau; **Cr** Plateau of Crapoel; **D** Plateau of Doenrade; **Gr** Plateau of Graetheide; **Hb** Heksenberg; **K** Plateau of Kerkrade; **M** Plateau of Margraten; **N** Plateau of Nieuwenhagen; **S** Plateau of Spekholzerheide; **U** Plateau of Ubachsberg; **V** Plateau of Vijlen; **A** Anstelerbeek; **C** Caumerbeek; **DC** Droogdal of Colmont; **E** Eyserbeek **G** Gerendal; **Gb** Geleenbeek; **Ge** Geul; **Gu** Gulp; **H** Basin of Heerlen; **J** Jeker; **Ma** Meuse; **R** Basin of the Roode Beek; **Se** Selzerbeek; **St** Strijthagerbeek; **W** Worm. **N.B.**: Beek = brook. Dal = valley, Berg = hill (literally meaning mountain) [Original Map by Romaine—CC0, *Source* Wikimedia Commons (2018b)]

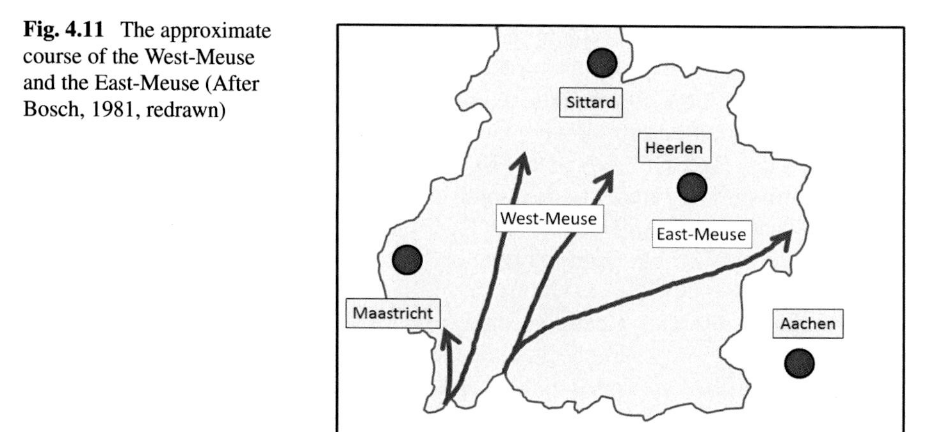

Fig. 4.11 The approximate course of the West-Meuse and the East-Meuse (After Bosch, 1981, redrawn)

Fig. 4.12 The subdivision of the Holocene (After the International Commission on Stratigraphy (2018h), and Encyclopedia Britannica (2018h), redrawn)

System / Period	Series / Epoch	Chronozone	Years Before Present
Quaternary	Holocene	Subatlantic	0 – 2.400
		Subboreal	2.400 – 5.660
		Atlantic	5.660 – 9.220
		Boreal	9.220 – 10.640
		Preboreal	10.640 – 11.650

ridge. Later, the Meuse broke through to the north. The valley of the East-Meuse has a width of maximum 6 km.

4.7.2 The Holocene

According to the International Commission on Stratigraphy the Holocene started approximately 11,700 years ago (ICS 2018g). The epoch follows the Pleistocene and the last glacial period (the Weichselian). The Holocene is subdivided into Preboreal, Boreal, Atlantic, Subboreal and Subatlantic (Fig. 4.12). Although large parts of The Netherlands became flooded, as a result of the melting ice-sheets, in the South Limburg area little happened. The tundra landscape of the ice ages disappeared, and plants and animals belonging to a more temperate climate zone became common. In the South-Limburg area, the main geological changes were the change of the course of the river Meuse, which altered its course from an easterly bedding to its present bedding. The present bedding runs from just north of Maastricht to the villages Maasbracht and Thorn, in the middle of Limburg, and through the course of that distance the Meuse constitutes the border between The Netherlands and Belgium. In the South Limburg area, during the period of this change of the course of the Meuse, sand and gravel were deposited locally (Geologie van Nederland 2017b).

4.8 The Coalmining Area

Having discussed the geological history of the South Limburg area, we will look now in detail at the coalmining area. The youngest rocks directly on top of the Carboniferous in the southeast are of Oligocene age, and in the extreme southeast of the South Limburg coal area (area of the Willem-Sophia and Domaniale Mines), the youngest cover on the Carboniferous rocks cover is even of Quaternary age. Near the Dutch town of Kerkrade, on the banks of the Worm (German: Wurm) River, a tributary of the Roer River, coal strata were even found lying at the surface, and have

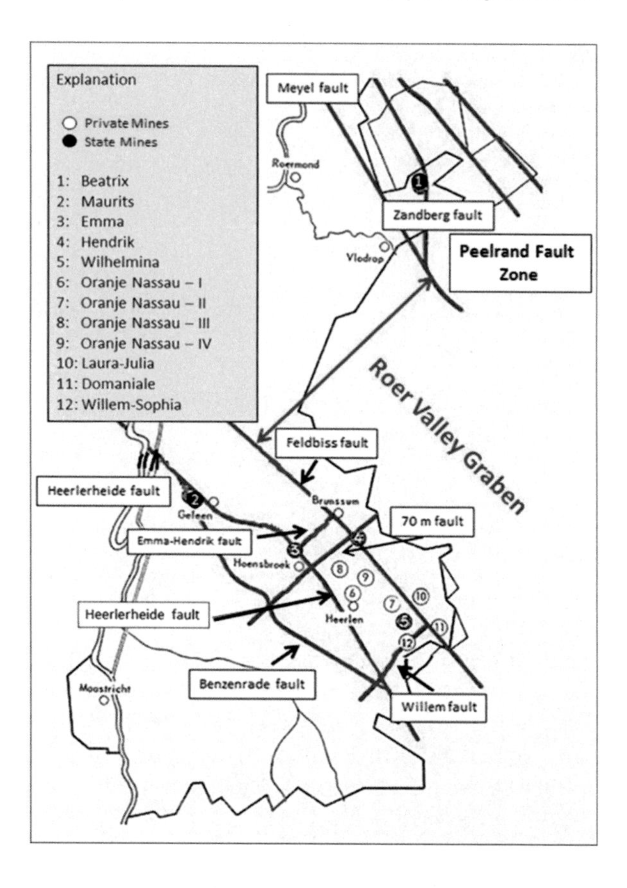

Fig. 4.13 Schematic picture showing the location of the coal mines with respect to the major faults in the area. SW–NE trending faults (Emma-Hendrik Fault, Willem Fault, and 70 m Fault) are of Variscan origin, and are not active anymore. Neither do they surface anywhere: they were only found in the coal deposits. (After Voncken and De Jong 2017, adapted)

been mined since medieval times (approximately 1100 AD, Voncken 2009). The fact that ever-younger rocks overlie the Carboniferous to the southeast is due to the dip of the Carboniferous to the northwest (Patijn and Kimpe 1961).

The present Roer Valley Graben is bordered in the South Limburg Area by the Peelrand fault zone to the northeast (consisting of the Peelrand, Meyel and Zandberg Faults) and the Feldbiss,[10] Heerlerheide, Geleen, Benzenrade and Kunrade Faults to the southwest. The main faults in the Dutch coal mining area are the Feldbiss Fault, the Benzenrade Fault and the Heerlerheide Fault. These faults have a NW-SE strike (Fig. 4.7 and 4.13). The throw in this fault system increases toward the NW from several meters to several hundreds of meters.

Below, a summary is given of the location of the mines with respect to the faults (Patijn and Kimpe 1961). Graphically this shown in Fig. 4.13.

Between the Benzenrade Fault and Heerlerheide Fault the following mines were situated:

[10]"Feldbiss" is a German word from the North Rhine-Westphalia region, which can best be translated as a "step in the field." Indeed, locally the fault can be observed as a step in the landscape (see for instance Houtgast 2018).

Oranje Nassau-III Mine

- Oranje Nassau-I Mine
- Emma State Mine
- Sophia-shaft of the Willem-Sophia Mine.

South of the Heerlerheide Fault, close to the joining of the Benzenrade Fault and the Heerlerheide Fault, only one mine was situated:

- Maurits State Mine.

Between Heerlerheide Fault and Feldbiss Fault the following mines were situated:

- Oranje Nassau-II Mine
- Oranje Nassau-IV Mine
- Hendrik State Mine (located just south of the Feldbiss fault, but also exploiting coal layers north of the fault. Shaft IV, depth 1058 m, was constructed to access the deeper situated coal layers north of the Feldbiss fault).
- Wilhelmina State Mine
- Laura Mine
- Willem-shaft of the Willem-Sophia Mine
- Neuprick Mine.

North of the Feldbiss Fault there were two mines:

- Julia Mine[11]
- Domaniale Mine.

Running more or less perpendicular to these faults, there are several older faults which do not surface. Some faults belonging to this older system are the "Willem Fault," the "Oranje Fault," the "70 m Fault", and the "Emma-Hendrik Fault." They are characterized by large differences in throw, in the Carboniferous, fluctuating from upthrows to downthrows, with occasionally very broad fault zones. They date from the times of Variscan (Hercynian) folding, and are not active any more (Patijn and Kimpe 1961). In the mines themselves, numerous small scale faults were discovered, which usually offset the coal layers by a distance of less than a meter to several meters.

It was also the intense faulting of the usually rather thin coal layers that made the underground mining in this area so costly, as extensive mechanized coal mining was not possible. However, mechanized coal mining did indeed occur, though usually carried out over only rather short distances.

[11]The Julia mine, owned by the same company as the Laura mine, was constructed because in the early 20th century, there was no technology available to penetrate the locally strongly water-bearing Feldbiss Fault. In later years, when technology had advanced enough, the two mines were connected by tunnels.

Folds are also encountered in the subsurface: the "Anticline of Waubach,"[12] the "Flexure (monocline) of Puth"[13] and the Flexure of Krawinkel.[14] These folds also have a Variscan origin (Patijn and Kimpe 1961). The strike of these folds is related to the upwarping of the Brabant Massif after the period of Variscan (Hercynian) folding. Overthrow tectonics due to Variscan (Hercynian) folding have been encountered almost exclusively in the most southern mines (Patijn and Kimpe 1961).

References

Bosch PW (1981) Het dal van de Oost-Maas in Zuid-Limburg. Grondboor en Hamer 3–4:95–109

De Mijnen (2018) http://www.demijnen.nl/actueel/artikel/bruinkoolwinning-eygelshoven-en-haanrade

Dinoloket (2018a) Triassic. https://www.dinoloket.nl/triassic. Visited May 2018

Dinoloket (2018b) Middle and lower jurassic. https://www.dinoloket.nl/middle-and-lower-jurassic. Visited May 2018

Dinoloket (2018c) Upper jurassic and lower cretaceous. https://www.dinoloket.nl/upper-jurassic-and-lower-cretaceous. Visited May 2018

Dinoloket (2018d) Upper cretaceous. https://www.dinoloket.nl/upper-cretaceous. Visited May 2018

Dinoloket (2018e) Kiezeloöliet formatie. https://www.dinoloket.nl/kiezelooliet-formatie. Visited May 2018

Doeglas DJ (1949) Loess, an eolian product. J Sed Res 19(3):112–117

Doppert JWChr, Ruegg GHJ, van Staalduinen CJ, Zagwijn WH and Zandstra JG (1975) Formaties van het Kwartair en Boven-Tertiair in Nederland. In: Zagwijn, WH & CJ van Staalduinen (red.), Toelichting bij geologische overzichtskaarten van Nederland. Rijks Geologische Dienst, Haarlem: 11–56 (In Dutch)

Duin EJT, Doornenbal JC, Rijkers RHB, Verbeek JW, Wong TE (2006) Subsurface structure of the Netherlands—results of recent onshore and offshore mapping. Netherlands J Geosci Geol Mijnbouw 85(4):245–276

Ellis D, Stoker MS (2014) The Faroe–Shetland Basin: a regional perspective from the Paleocene to the present day and its relationship to the opening of the North Atlantic Ocean. In: Cannon SJC, Ellis D (eds) Hydrocarbon exploration to exploitation west of shetlands, vol 397. Geological Society, London, Special Publications, pp 11–31

Encyclopedia Britannica (2018a) Permian. https://www.britannica.com/science/Permian-Period. Visited Mar 2018

Encyclopedia Britannica (2018b) Triassic. https://www.britannica.com/science/Triassic-Period/Economic-significance-of-Triassic-deposits. Visited Mar 2018

Encyclopedia Britannica (2018c) Jurassic. https://www.britannica.com/science/Jurassic-Period/Jurassic-geology#ref257903. Visited Apr 2018

Encyclopedia Britannica (2018d) Cretaceous. https://www.britannica.com/science/Cretaceous-Period/Major-subdivisions-of-the-Cretaceous-System. Visited Apr 2018

Encyclopedia Britannica (2018e) Paleogene. https://www.britannica.com/science/Paleogene-Period. Visited May 2018

[12]Waubach is a part of the village Ubach over Worms, presently part of the municipality of Landgraaf.

[13]Puth is a little township belonging to the village of Schinnen, itself situated some km's east of Geleen-Sittard.

[14]Krawinkel is a township close to Geleen-Sittard.

Encyclopedia Britannica (2018f) Neogene. https://www.britannica.com/science/Neogene-Period. Visited May 2018

Encyclopedia Britannica (2018g) Pleistocene period. https://www.britannica.com/science/Pleistocene-Epoch. Visited May 2018

Encyclopedia Britannica (2018h) Holocene period. https://www.britannica.com/science/Holocene-Epoch. Visited May 2018

Felder WM, Bosch PW (2000) Krijt van Zuid-Limburg. Geologie van Nederland, part 5, Instituut voor Toegepaste Geowetenschappen TNO, Delft/Utrecht, 190 pp (In Dutch)

de Gans W (2007) Quaternary. In: Wong ThE, Batjes DAJ, de Jager J (eds). Geology of the Netherlands. KNAW

Geluk MC, Duin EJTh, Dusar M, Rijkers RHB, van den Berg MW, Van Rooijen P (1994) Stratigraphy and tectonics of the Roer Valley Graben. Geol Mijnbouw 73:129–141

Geologie van Nederland (2017a) http://www.geologievannederland.nl/tijd/reconstructies-tijdvakken/mioceen. Visited Nov 2017

Geologie van Nederland (2017b) http://www.geologievannederland.nl/tijd/reconstructies-tijdvakken/holoceen#head6. Visited Nov 2017

Heerlen Vertelt (2018) http://www.heerlenvertelt.nl/2011/02/heerlen-toen-steenkool-en-bruinkool

Houtgast R (2018) Photographs of the Feldbiss Fault. https://www.geo.vu.nl/~balr/hour/foto's1.htm

ICS Holocene (2018h) http://quaternary.stratigraphy.org/workinggroups/holocene/. Visited May 2018

ICS (2018a) Permian. http://permian.stratigraphy.org/per/per.asp. Visited Mar 2018

ICS (2018b) Triassic. https://albertiana-sts.org/. Visited Mar 2018

ICS Jurassic (2018c) https://jurassicdotstratigraphydotorg.files.wordpress.com/2015/12/timescalecreators_jurassic.pdf. Visited Apr 2018

ICS Cretaceous (2018d) http://cretaceous.stratigraphy.org/. Visited Apr 2018

ICS Paleogene (2018e) http://www.paleogene.org/. Visited May 2018

ICS Neogene (2018f) http://www.sns.unipr.it/. Visited May 2018

ICS Pleistocene (2018g) http://quaternary.stratigraphy.org/definitions/pleistocenesubdivision/. Visited May 2018

Imeson A (2012) Desertification, land degradation and sustainability, Chap. 6, p 209. Wiley-Blackwell Publishers, 325 pp

Michon L, Van Balen RT, Merle O, Pagnier H (2003) The Cenozoic evolution of the Roer Valley Rift System integrated at a European scale. Tectonophysics 367:101–126

Patijn RJH, Kimpe WFM (1961) De kaart van het Carboon oppervlak, de profielen en de kaart van het dekterrrein van het Zuid-Limburgs Mijngebied, en Staatsmijn Beatrix en omgeving. Mededelingen van de Geologische Stichting, Serie C, 1—1—No. 4, 1961, publishing company "Ernest van Aelst, Maastricht" (With Dutch and English description)

Ritsema AR (1985) Intraplate Seismic Activity in the Brabant Massif and the graben systems of the lower Rhine-Roer and North Sea. In: Seismic Activity in Western Europe, with particular consideration to the Liège Earthquake of November 8, 1983, P. Melchior editor. NATO ASI Series C, Mathematical and Physical Sciences, D. Reidel Publishing Company, Dordrecht/Boston/Lancaster, vol 144, pp 85–99

Voncken JHL (2009) Ontstaansgeschiedenis van de steenkolenwinning in Nederland. GEA 42(4):112–115 (In Dutch)

Voncken JHL, de Jong TPR (2017) Coal mining in the Netherlands; http://www.citg.tudelft.nl/en/about-faculty/departments/geoscience-engineering/related-links/coal-mining-in-the-netherlands/coal-data/geology-and-coal/. Visited Mar 2018

Westerhof WE (2003) Nomenclator Formatie van Breda. Dutch Institute for Applied Geosciences—TNO, Utrecht, 7 pp

Wikimedia Commons (2018a) Roerdal Graben https://commons.wikimedia.org/wiki/File:Roerdal_graben_map_NL.svg. Visited Mar 2018

Wikimedia Commons (2018b) Map with plateaus and valleys of South-Limburg. https://commons.wikimedia.org/wiki/File:Kaart_met_plateaus_en_dalen_Zuid-Limburg.PNG. Visited Apr 2018 (Redrawn)

Wikipedia (2018a) Zilverzand. https://nl.wikipedia.org/wiki/Zilverzand. Visited May 2018 (In Dutch)

Wikipedia (2018b) Loess. https://en.wikipedia.org/wiki/Loess. Visited Feb 2018

Zagwijn WH (1989) The Netherlands during the Tertiary and the Quaternary: a case history of coastal lowland evolution. Geol Mijnbouw 68:107–120

Ziegler PA (1989) Evolution of Laurussia: a study in Late Palaeozoic plate tectonics. Kluwer Academic Publishers, 115 pp

Ziegler PA (1992) North Sea rift system. Tectonophysics 208:55–75

Printed in the United States
By Bookmasters